REACTIVE
FREE
RADICALS

REACTIVE
FREE
RADICALS

J. M. HAY

Vientiane,
Laos

1974

ACADEMIC PRESS
London and New York

A Subsidiary of Harcourt Brace Jovanovich, Publishers

ACADEMIC PRESS INC. (LONDON) LTD.
24/28 Oval Road
London NW1

United States Edition published by
ACADEMIC PRESS INC.
111 Fifth Avenue
New York, New York 10003

Library of Congress Catalog Card Number: 73-19007
ISBN: 0 12 333550 7

Set in 'Monophoto' Times and printed in Great Britain by
Page Bros (Norwich) Ltd, Norwich

PREFACE

Authors of most books on the chemistry of free radicals tend to adopt the phenomenological approach to their subject, describing and classifying the preparations, properties and reactions of radicals without a main unifying and linking theme other than the obvious one of the presence in each of these species of at least one unpaired electron. The present book is a tentative step based on very elementary, perhaps even naive, views of the nature of radicals towards a structure-reactivity relationship which can be applied to these species.

The initial premise is that the reactivity of a free radical depends upon its shape and therefore upon the character of the orbital(s) containing the unpaired electron. This is not a new idea but previous authors have tended to bring it in as a sidewind to other more important arguments. Here it is developed as the main theme and shown to be a possible explanation for certain of the fascinating phenomena observed in a number of free radical reactions such as combustion and pyrolysis. It also provides, by showing that in a radical reaction the shapes of the reactants and therefore the kinetics of the reaction may not be invariant under different experimental conditions, a plausible solution to one of the major difficulties hindering the development of a comprehensive theory of free radical reactions, that of the lack of agreement between the experimental results of investigators of apparently identical reactions. The theory is also extended to provide information on the thermodynamics of organic species and, briefly, on the structure of radicals and polar effects in their reactions.

CONTENTS

CHAPTER 1

INTRODUCTION

This chapter takes the form of a brief résumé of the chemistry of free radicals, what they are and how they are formed, their structure and reactivity. The subsequent chapters add the fine detail to this broad-brush background presenting ultimately, it is hoped, a picture which will explain the continued attraction of this branch of chemistry for so many practical and theoretical chemists. The treatment is fairly general but concentrates on those character-istics of free radicals attributable directly to the unpaired electron function. Omitted altogether are the non-radical reactions of free radicals which have been described in a recent review.[1]

For the purposes of this book, a free radical is a molecule, molecular fragment, complex or atom which incorporates one or more unpaired electrons and the properties of which are controlled by such electron(s). The stability of radicals varies from those like diphenylpicrylhydrazyl (DPPH) (I),

(I)

which can be synthesized, isolated and crystallized using conventional techniques of organic chemistry and kept indefinitely in a bottle (although reaction with oxygen, which is also a free radical, may cause gradual de-terioration) through those which can be isolated for sufficiently long to be studied by conventional methods of chemical spectroscopy, to those which are so short-lived that, unless trapped in an inert solid matrix, require for their study specially devised techniques. Broadly, the chemical and physical properties of the first two types of radical are covered in ref. 1. While they will not be altogether ignored in the present book, we are here more concerned with the most reactive radicals.

The study of reactive radical reactions should ideally involve the generation of the species of interest and their monitoring *in situ*. Suitable techniques are described extensively in the chemical literature but the chemical problems associated with the identification and isolation of the desired reactions are very much a function of the system which is being studied. Although these difficulties will not be considered *per se*, their existence will become clear during the course of this and the following chapters.

THE FORMATION OF FREE RADICALS

The formation of free radicals from neutral molecules requires energy in the form of radiation (heat, u.v., microwave, X-ray, electrons, α-particles, neutrons, γ-rays) or from electron transfer or mechanical degradation. Charged radicals are normally produced by electron transfer at electrodes or from charged species. A whole spectrum of radicals can be obtained from one organic molecule with the input of a sufficient dose of energy as exemplified nicely by the detection of the range of molecular fragment ions (and associated radicals) produced from ethane upon high energy electron bombardment in the mass spectrometer:[2]

$$C_2H_6 + ne \rightarrow C_2H_6^+, C_2H_5^+, C_2H_4^+, C_2H_3^+, C_2H_2^+, C_2H^+, C_2^+,$$
$$CH_3^+, CH_2^+, CH^+, C^+,$$
$$H_2^+, H^+,$$
$$C_2H_5^{\cdot}, CH_3^{\cdot}, C_2H_3^{\cdot} \text{ and } H^{\cdot}.$$

Such an embarrassment of products is typical of high energy irradiation techniques and to obtain detailed results amenable to conventional treatment, radical systems are kept as simple as possible.

The use of visible or u.v. photons with suitable filters often allows a choice of those frequencies which are absorbed by the relevant part or parts of the molecule. Some typical photolytic sources of free radicals are,

$$Cl_2 \xrightarrow{h\nu} 2Cl^{\cdot}$$
$$RCOR' \xrightarrow{h\nu} RCO^{\cdot} + R''$$
$$RI \xrightarrow{h\nu} R^{\cdot} + I^{\cdot}$$

Where simple photolysis is inconvenient, atom sensitization may be tried, as exemplified by the mercury sensitized decomposition of a hydrocarbon,

$$Hg(^1S_0) + h\nu(2537 \text{ Å}) \rightarrow Hg(^3P_1)$$

$$Hg(^3P_1)(112\cdot7 \text{ kcal}) + RH(^1\Sigma) \rightarrow Hg(^1S_0) + R^{\cdot}(^2\Sigma) + H(^2S)$$

Another approach is to work at relatively low temperatures and to this end a large number of thermal sources of free radicals has been devised. Typical of this class of compounds are the peroxides,

ΔH (kcal mol^{-1})

Hydroperoxides	$ROOH \rightarrow RO^{\cdot} + {}^{\cdot}OH$	$c.$ 38
Dialkylperoxides	$ROOR' \rightarrow RO^{\cdot} + {}^{\cdot}OR'$	$c.$ 35
Peracids	$RCOOOH \rightarrow RCO_2^{\cdot} + {}^{\cdot}OH$	30
Peresters	$RCOOOR' \rightarrow RCO_2^{\cdot} + {}^{\cdot}OR'$	20–30

Peroxydicarbonates

$$\underset{ROCOOCOR'}{O\ \ \ \ O} \rightarrow \underset{ROCO^{\cdot}}{O} + \underset{R'OCO^{\cdot}}{O}$$
$$\downarrow$$
$$RO^{\cdot} + CO_2$$

Diacylperoxides	$RCOOOCOR' \rightarrow 2RCO_2^{\cdot}$	$c.$ 30
Persulphate	$O_3S\bar{O}{-}OSO_3^{-} \rightarrow 2SO_4^{\cdot}$	33·5

and certain azo-compounds, e.g.

Di-t-butylhyponitrite
$(CH_3)_3CO{-}N{=}N{-}OC(CH_3)_3 \rightarrow (CH_3)_3CO^{\cdot} + N_2 + (CH_3)_3CO^{\cdot}$ 27·9

Azoisobutyronitrile

$$\underset{CN}{(CH_3)_2C}{-}N{=}N{-}\underset{CN}{C(CH_3)_2} \rightarrow 2\underset{CN}{(CH_3)_2C^{\cdot}} + N_2 \qquad 30{\cdot}8$$

Azoisopropane
$(CH_3)_2CHN{=}NCH(CH_3)_2 \rightarrow 2(CH_3)_2CH^{\cdot} + N_2$ 40·9

At high temperatures the potential sources of free radicals are numerous but special mention may be made of metal alkyls, i.e.

$$R_2Hg \rightarrow RHg^{\cdot} + R^{\cdot} \qquad\qquad 40\text{–}50$$

and alkyl nitrites $RONO \rightarrow RO^{\cdot} + NO$ 35–40

In polar solutions redox systems provide a convenient and prolific source of radicals, e.g.

$$RCO_2^{-} - e \rightarrow RCO_2^{\cdot}$$
$$ROOH + Fe^{2+} \rightarrow RO^{\cdot} + Fe(OH)^{2+}$$
$$\rightarrow RO^{\cdot} + Fe^{3+} + OH^{-}$$

Many other sources of radicals have been contrived, three only of which will be noted particularly although others will appear below:

(a) The addition of hydrogen atoms to unsaturated molecules,

$$H_2 \rightarrow 2H^{\cdot}$$

$$H^{\cdot} + A = B \rightarrow HA - B^{\cdot}$$

Experimentally the hydrogen atoms are produced in a stream of hydrogen gas and the olefin or acetylene is introduced downstream.

(b) The reverse of disproportionation, e.g.

$$RH + O_2 \rightarrow R^{\cdot} + HO_2^{\cdot} \qquad \Delta H = 40 \, kcal \, mol^{-1}$$

and possibly,

$$C_2H_6 + C_2H_4 \rightarrow 2C_2H_5^{\cdot} \qquad \Delta H = c. \, 60 \, kcal \, mol^{-1}$$

the latter process having a considerably smaller endothermicity than the formation of small free radicals from ethane by the rupture of a C—C or C—H bond, i.e.

$$C_2H_6 \rightarrow 2CH_3^{\cdot} \qquad \Delta H = 86 \, kcal \, mol^{-1}$$

$$\rightarrow C_2H_5^{\cdot} + H^{\cdot} \qquad \Delta H = 98 \, kcal \, mol^{-1}$$

(c) Ionic disproportionation, e.g.

$$ArO^+ + ArO^- \rightleftharpoons 2ArO^{\cdot}$$

THE FORMATION OF RADICALS BY UNIMOLECULAR DISSOCIATION

We will now consider in slightly more detail the mechanisms of the formation of free radicals beginning with the simplest, the dissociation of a chemical bond. The rate of a unimolecular reaction such as a molecular dissociation is dependent not only upon its activation energy but also on the entropy change associated with the process, the inter-relationship between the two terms being expressed in the simple transition state equation for the rate constant (1),

$$k = \frac{\kappa kT}{h} \exp\left(\Delta S^{\ddagger}/R\right) \exp\left(-\Delta H^{\ddagger}/RT\right) \qquad (1)$$

where k is Boltzmann's constant, h is Planck's constant and κ is the transmission coefficient, usually and for convenience set equal to unity. $\Delta S^{\ddagger}$ and $\Delta H^{\ddagger}$ are respectively the changes in entropy and enthalpy associated with the formation of the activated complex, the latter being related to the potential energy barrier for the reaction. From the entropy term it can be seen that if the transition state is looser than the original molecule, i.e. the dissociating parts are freer to vibrate and rotate, the A-factor, as measured from the Arrhenius equation,

$$k = A \exp(-E/RT) \tag{2}$$

is greater than kT/h ($c.\ 4 \times 10^{13}\ s^{-1}$ at room temperature). Conversely, if the formation of the transition state involves a tightening of the molecule then A becomes less than $10^{13}\ s^{-1}$.

Whether or not the decomposition of a molecule follows unimolecular kinetics depends on the rate at which the energy of activation is transmitted to the molecule and eventually to the bond which is breaking allowing for the competing process of deactivation by collision with other species. Since, in the absence of incident radiation, the energy is attained by bimolecular collision, at sufficiently low pressure where the probability of deactivation by collision is low, the rate of the reaction will be controlled by the rate of this energy acquisition and the kinetics will be second order. As the pressure is increased collisional deactivation increases in importance and as its rate overtakes that of the unimolecular decomposition, the latter becomes rate determining and the reaction becomes first order, the first order rate constant increasing to a limiting high pressure value.

The main theoretical approaches to this problem have had some success. One, the HKRR, associated with the names of Hinshelwood, Kassel, Rice and Ramsperger and further developed by Marcus (the HKRRM theory)[3] is based on the assumption that any energy acquired by the molecule can readily be distributed amongst its various modes of excitation, and reaction occurs when the bond to be broken accumulates the necessary and sufficient amount of energy. The other, developed by Polanyi and Wigner and ultimately by Slater[4] in its extreme form permits no energy exchange between the normal modes of the molecule; instead reaction occurs when that combination of normal mode vibrations occurs which brings the critical coordinate, in our case the bond to be broken, above a certain limiting extension. Certain imperfections of this theory which have become apparent in its application to a number of systems have stimulated its extension to incorporate a degree of intramolecular energy transfer.

From the HKRR theory, the pressure below which the unimolecular rate constant drops off can be shown to depend upon the number of degrees of freedom, s ($= 3n - 6$ where n is the number of atoms) in which the energy can be spread. Within the molecule there are many dispositions of the energy above the minimum, E^*, at which reaction can occur but the probability that E^* accumulates in the bond under consideration is $(1 - E/E^*)^{s-1}$ where E is the total energy content of the reacting molecule, whence the rate constant for the dissociation of the molecule is,

$$k_{E^*} = v(1 - E/E^*)^{s-1} \tag{3}$$

v being a proportionality factor related to the rate of internal energy transfer.

Since $E^* \geqslant E$ the greater the number of degrees of freedom, i.e. the larger that part of the molecule containing the system of coupled oscillators, the greater is the value of k_{E^*} and the lower the pressure region at which fall-off occurs. We therefore expect to find fall-off at experimentally conveniently attainable pressures either when the molecule is small or when there is inefficient intramolecular energy transfer. The expected ranges of pressure at which fall-off becomes appreciable are:[5]

s	p (mmHg)
4	3300
8	24
12	0·91
15	0·14
21	0·0091

At sufficiently low pressures eventually the reaction becomes second order, controlled by a limiting low pressure rate constant,

$$k = \frac{Z}{(s-1)!}(E/RT)^{s-1}\exp{(-E/RT)} \tag{4}$$

where Z is the collision number.

Much ingenuity has been expended in attempts to find reactions which can be used to test these various theories. The most straightforward are those which are chemically "clean" and uncomplicated by secondary reactions. Certain cycloalkane and alkene isomerizations fill this role admirably.[6] The first such system to be treated theoretically and experimentally was the isomerization of cyclopropane,

$$\triangle \rightarrow CH_3CH{=}CH_2$$

Slater,[4] taking as his critical coordinate the distance between a carbon atom and a hydrogen atom attached to an adjacent carbon atom was able to match theoretical prediction to experimental finding with uncanny precision.

Other examples of reactions which have been so studied are,[7, 8]

$$CH_3NC \rightarrow CH_3CN$$

$$(CH_2O)_3 \rightarrow 3CH_2O$$

The decomposition of neutral molecules into radicals is rather more difficult to study since the products are likely to attack the parent molecule in secondary processes which will tend to swamp the reaction under examination especially if these secondary processes give rise to a chain reaction. One way round this problem is to make use of the technique of radical trapping which effectively removes the product radicals in a more rapid process than their attack on the parent molecule, for example,

$$R\!-\!R' \rightarrow R^{\bullet} + R''^{\bullet}$$

$$R^{\bullet}(R'') + HI \rightarrow RH(R'H) + I^{\bullet}$$

where $I^{\bullet}$ is sufficiently inert not to stimulate the formation of a new radical $R^{\bullet}$ or in other words, the rate of the reaction,

$$RH(R'H) + I^{\bullet} \rightarrow R^{\bullet}(R'') + HI$$

is negligibly slow under the experimental conditions employed. But the addition of a second component to the reaction mixture is obviously a disadvantage when low pressures are required.

Kinetic analysis of complex reaction systems is also possible but often difficult. A good example of a fairly complicated system which is amenable to experimental and theoretical treatment is the decomposition of the alkoxy radical produced in the pyrolysis of a peroxide[9] or alkyl nitrite:[10]

$$ROOR' \rightarrow RO^{\bullet} + R'O^{\bullet}$$

$$RONO \rightarrow RO^{\bullet} + NO$$

followed by,

$$RO^{\bullet} \rightarrow carbonyl + radical$$

e.g.

$$(CH_3)_2CHO^{\bullet} \rightarrow CH_3CHO + CH_3^{\bullet}$$

If it can be shown that the carbonyl has no source other than the decomposition of the alkoxy radical, then chemical analysis of the reaction mixture for its kinetics of formation with variation of pressure will show up any fall-off behaviour.

A common approach is to study the reverse reaction, i.e.

$$R^{\bullet} + R''^{\bullet} \rightarrow RR'$$

Provided that the bimolecular reaction occurs by the reverse of the unimolecular (the principle of microscopic reversibility) the kinetics of the forward reaction can be calculated and treated theoretically—a necessary approach since bimolecular reactions are not yet amenable to detailed theoretical treatment. In this way a good deal of information has been obtained concerning the recombination of atoms produced in shock tubes.

by flash photolysis and by discharge methods. In the bimolecular association of atoms there is no way in which the energy of the resulting bond can be dissipated within the product molecule which flies apart immediately unless, by means of a collision with a third body, the energy can be reduced below that for dissociation of the bond. Studies of atom recombination rates in the presence of different inert additives have provided much information about energy transfer mechanisms and have shown that both light and heavy molecules can act as efficient third bodies, the former carrying the excess energy away in increased translational energy and the latter in the various degrees of freedom.

Sometimes a highly specific effect is noted; for example it seems likely that resonance between the vibration frequencies in the bond formed and in the third body can provide an efficient pathway for stabilization, as for example in the combination,[11]

$$H^{\cdot} + O_2 + M \rightarrow HO_2^{\cdot} + M$$

Stronger effects still, giving very high collision efficiencies, are observed when a complex is formed with the additive, e.g.

$$ArH + I^{\cdot} \rightarrow (IArH)^{\cdot}$$

$$^{\cdot}IArH + I^{\cdot} \rightarrow I_2 + ArH$$

The formation of complexes between iodine atoms and aromatics has been demonstrated using u.v. spectroscopy.[12]

Kinetic analysis has also given information about the pressure dependence of bimolecular collisions between radicals; such reactions between methyl radicals and between methyl and oxygen and nitric oxide have been found to exhibit such effects under normal conditions.[13]

The flow and distribution of energy in molecules can be observed in experiments in which exothermic bimolecular combination reactions are followed by unimolecular decompositions. Particularly interesting are those cases where the unimolecular decomposition can proceed by different pathways to give alternative sets of products, e.g.

$$A + B \rightleftharpoons AB^*$$

$$AB^* + M \rightarrow AB + M$$

$$AB^* \rightarrow C + D$$

$$\rightarrow E + F$$

This method of energizing molecules or radicals has the advantage that, provided the original species A and B are in thermal equilibrium in the system, the amount of excess energy, the dissociation energy of the bond being formed, and its location, the site of the new bond, are known. From the products of the reaction the probabilities of the alternative decomposition pathways and therefore of the different distributions of the excess energy can be deduced.[14]

Reactions studied in the liquid phase are in the main free from difficulties caused by excess energy since the close contact between molecules in the condensed phase facilitates rapid energy dissipation. The liquid phase is also free (but see ref. 15) from the kind of wall and other heterogeneous effects which have lured many a gas phase experimenter into a misguided match of experimental result and simple theory. The very much lower rates of diffusion in the liquid phase do, however, pose some peculiar problems. In the first place the radicals, $R^{\cdot}$ and R'', formed by thermal or photochemical dissociation,

$$R{-}R' \to R^{\cdot} + R''$$

are initially separated by a distance presumably not much greater than that of the $R{-}R'$ bond and are surrounded by a "cage" of solvent molecules which, in preventing the radicals from moving apart, makes it likely that they will come together and recombine. The quantum yield of radicals, as measured for example by the quantity of radiant energy absorbed, is thus reduced. It is usually easy to compensate for this cage effect by measuring the rate of removal of the parent molecule which will give the net rate of formation of radicals. But complications can arise if the cage recombination does not reform the reactant molecule as probably occurs in the dissociation of azoisobutyronitrile,

$$\underset{\substack{| \\ (CH_3)_2C}}{\overset{CN}{|}}N{=}\underset{\substack{| \\ C(CH_3)_2}}{\overset{CN}{|}} \to \underset{\substack{| \\ (CH_3)_2C}}{\overset{CN}{|}}N_2^{\cdot} + \underset{\substack{| \\ {}^{\cdot}C(CH_3)_2}}{\overset{CN}{|}}$$

$$N_2 + \underset{\substack{| \\ {}^{\cdot}C(CH_3)_2}}{\overset{CN}{|}}$$

$$\underset{\substack{| \\ (CH_3)_2C}}{\overset{CN}{|}}{-}\underset{\substack{| \\ C(CH_3)_2}}{\overset{CN}{|}}$$

The resulting dinitrile does not absorb at the same wavelength as does the reactant so that measurement of the rate of disappearance of the azoiso-butyronitrile by monitoring its u.v. absorption does not give a true measure of the net rate of formation of radicals. Electron spin resonance spectroscopy (e.s.r.) promises to outflank at least some of these difficulties since the lifetime of a radical in a cage being less than about 10^{-6} s, is too short for detection by conventional e.s.r. techniques.

Outside of the cage radical–radical reactions normally occur at rates which are faster than those at which the radicals can diffuse together and are therefore diffusion controlled with a rate constant under steady–state diffusion conditions of,[16]

$$k = \frac{4\pi r_{AB}D_{AB}}{1 + (4\pi r_{AB}D_{AB})/k_B} \quad \text{molecules cm}^{-3}\,\text{s}^{-1} \tag{5}$$

where r_{AB} is the radius of the "capture" sphere formed by B molecules around an A molecule, D_{AB} is the relative diffusion coefficient of the reactants, equal to the sum of the individual diffusion coefficients, D_A and D_B, and k_B is the rate constant which would be observed if the concentration of B molecules in the nearest neighbour shell of an A molecule were the same as the average over the whole system.

A further potential complicating feature of solution free radical chemistry is the possibility of attack on the solvent by radicals and the consequent production of secondary radicals, the reactions of which may not easily be taken into account in the kinetic analysis.

Finally there is the question of radical–solvent interactions such as occur for example between chlorine atoms and benzene to give a complex,

thus increasing both the selectivity and the electronegativity of the chlorine atoms. E.s.r. is the obvious tool with which to probe radical–solvent interactions since the shapes of radical spectra are delicately sensitive to electron density. For simple radicals in non-polar solvents these effects are small but they do become noticeable as radical and solvent polarities increase. For example the equilibrium between p-phenylene diamine and chloranil,

lies well to the left in non-polar solvents but as the dielectric constant of the solution is increased it shifts to the right and the radical ion e.s.r. spectra become detectable.[17]

Polar effects are also observed in the radical spectra from nitroso compounds, semiquinones, thio-indigo and the neutral nitroxides.[1] The consequent changes in proton e.s.r. hyperfine splittings are not in general very large but those of ^{14}N and ^{13}C, which are more sensitive to the detailed distribution of the free electron can be altered quite dramatically. The effects are attributed to the formation of solvent–radical complexes which undergo exchange, which can be fast or slow, and which cause a redistribution of the π-electron density. Clearly there is considerable scope for parallel study on radical reactivity and free electron distribution.

In the great majority of cases, however, reactions of radicals in the gas phase and solution phases are closely similar and medium effects on the reactivity of neutral radicals insigificant.

TYPICAL REACTIONS OF FREE RADICALS

We shall now turn to the typical reactions of free radicals and survey briefly their mechanistic characteristics rather than their detailed kinetic features which will be considered in Chapter 4.

A. Unimolecular Reactions:
   (*i*)   Isomerization $R^{\cdot} \rightarrow R_1^{\cdot\prime\prime}$
   (*ii*) Decomposition $R^{\cdot} \rightarrow R_2^{\cdot\prime\prime}$ + products

B. Bimolecular reactions:
   (*i*)   Combination $R^{\cdot} + R^{\prime\prime} \rightarrow RR^{\prime}$
   (*ii*)   Disproportionation $R^{\cdot} + R^{\prime\prime} \rightarrow RH + R^{\prime}(-H)$
   (*iii*) Transfer $R^{\cdot} + R^{\prime}X \rightarrow RX + R^{\prime\prime}$
   (*iv*) Addition $R^{\cdot} + X{=}Y \rightarrow RXY^{\cdot}$
   (*v*)   Electron transfer $R^{\cdot} + R^{\prime+\cdot} \rightarrow R^{+\cdot} + R^{\prime\prime}$
                                  $R^{\cdot} + R^{\prime-\cdot} \rightarrow R^{-\cdot} + R^{\prime\prime}$
                                  $R^{\cdot} + R^{\prime\prime} \rightarrow R^{-\cdot} + R^{\prime+\cdot}$
   (*vi*) Aromatic substitution $R^{\cdot} + ArY \rightarrow RAr + Y^{\cdot}$
   (*vii*) Displacement $R^{\cdot} + R^{\prime}Y \rightarrow RR^{\prime} + Y^{\cdot}$

Many unimolecular reactions are merely the intramolecular analogues of those under section B.

A. Unimolecular Reactions

(*i*) *Isomerization*

Radical isomerizations are roughly of two kinds, those in which an atom or group migrates and those in which the skeleton of the radical rearranges. In the former, the site bearing the unpaired electron is supposed to approach to within "reaction distance" of the moving group and the energetics and entropy of the process are controlled by the strain energy of the ring formed in the transition state (II),

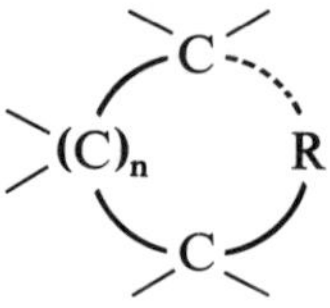

(II)

For a hydrogen atom transfer the activation energy of the process is that normally associated with a hydrogen atom transfer (see below) augmented by the ring strain energy which has been given approximately for saturated cycloalkane rings of three members as 28 kcal mol^{-1}, of four members as 26 kcal mol^{-1}, of five members as 6·5 kcal mol^{-1}, of six members as 0·6 kcal mol^{-1} and of seven members as 6·5 kcal mol^{-1}.[18] 1,2 shifts involving a three membered ring transition state have been examined theoretically by Phelan et al.[19] who have shown that phenyl shifts more easily than methyl and that hydrogen atom 1,2 shifts are the most difficult of all.

Experimentally, the occurrence of 1,2 hydrogen atom shifts has been doubted but other workers have proposed the following reactions,

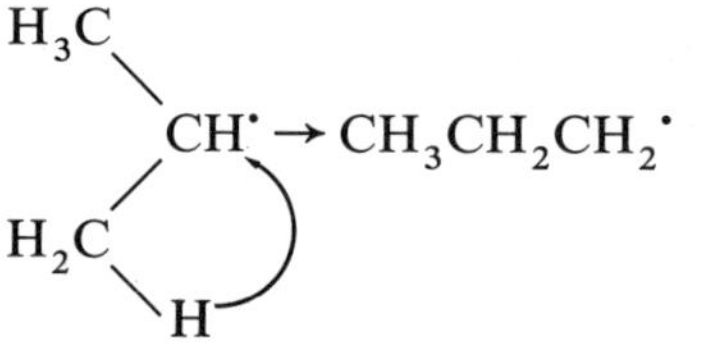

during the decomposition of the isopropyl radical[20] and the photolysis of di-isopropyl ketone in the presence of metals. Frey and Walsh[21] report that the A-factor for the reaction is "normal" at about $10^{13 \pm 1}$ and the activation energy, calculated, is $43 \pm 3·5$ kcal mol^{-1}. A similar reaction,

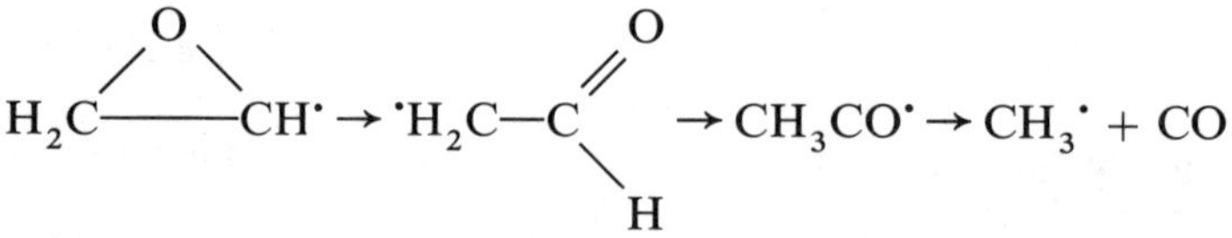

has been proposed[22] in the decomposition of ethylene oxide. On the other hand no 1,2 hydrogen atom shift was detected in the ethyl radical under conditions in which a 1,3 shift was detected in the n-propyl radical:[23]

$$CH_3CH_2CH_2^{\cdot} \rightarrow {}^{\cdot}CH_2CH_2CH_3$$

Benson[24] has doubted the proposition of a 1,3 shift during the oxidation of methyl,

$$CH_3^{\cdot} + O_2 \rightarrow CH_3O_2^{\cdot} \rightarrow {}^{\cdot}CH_2OOH \rightarrow CH_2O + {}^{\cdot}OH$$

despite the fact that the peroxy radical is excited to the extent of some 17 kcal mol^{-1} originally mainly located in the $C-O$ bond formed in the initial association. We shall revert to this reaction in Chapter 4.

The kinetic parameters of the 1,4 hydrogen atom shift,

$$CH_3CH_2CH_2CH_2CH_2^{\cdot} \longrightarrow CH_3\overset{\cdot}{C}HCH_2CH_2CH_3$$

have been given[25] as $A = 10^{7.15}$ and $E = 10.8$ kcal/mol^{-1}, the former being rather low but the latter falling into the correct pattern for a hydrogen atom transfer. 1,4, 1,5 and 1,6 atom transfers are all common, for example in the isomerization of certain peroxy radicals,[26]

This type of reaction can also occur in polymerization, for example of ethylene where the formation of numbers of 4-membered branches can be traced to an intramolecular hydrogen atom transfer.[27] The length of the branch reflects the ease of formation of 6-membered rings,

The transfer of hydrogen atoms also occurs in aromatic systems, for example,[28, 29]

and perhaps the most puzzling reaction of this type,[30]

$$k = 3 \cdot 1 \times 10^6 \exp(-11 \cdot 4/RT)$$

the unusually low A-factor of which raises some doubts as to its unimolecularity. But Bennett has argued that if such an intramolecular rearrangement were to go by way of a π complex, the extra electron of the free radical would make the formation of the complex more difficult and therefore the A-factor even lower.

More unusual is the migration of groups other than hydrogen atoms but the formation of the final products from the decomposition of β-methyl, β-phenyl, β-peroxypropiolactone appears to involve a methyl shift,[31]

Migration of an aryl group is presumed to occur in the decomposition of triphenylmethyl peroxide,[32]

A further interesting example of unimolecular radical isomerization is the intramolecular addition to form a nitroxide radical,[33]

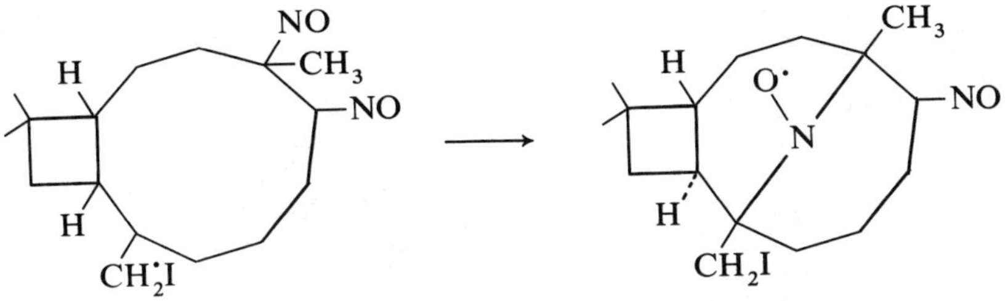

(ii) Decomposition

The decomposition of radicals is usually driven by the energy produced in the formation of an unsaturated bond. For example the reaction,

$$CH_3CH_2CH_2^\cdot \longrightarrow CH_3^\cdot + CH_2{=}CH_2 \ \Delta H = +24 \, kcal$$

requires much less energy than the dissociation of propane, the lowest energy path of which also involves a $C-C$ bond rupture, and the energetics are even more favourable where the decomposition involves the formation of a strongly bonded neutral molecule such as oxygen or one with a carbonyl bond,

$$CH_3\dot{C}O \rightarrow CH_3^\cdot + CO \qquad \Delta H = c. \ 12 \, kcal$$
$$CH_3\dot{C}O_2 \rightarrow CH_3^\cdot + CO_2 \qquad \Delta H = -20 \, kcal$$
$$(CH_3)_3CO^\cdot \rightarrow (CH_3)_2CO + CH_3^\cdot \qquad \Delta H = 7 \, kcal$$
$$CH_2{=}CHCH_2{-}O_2^\cdot \rightarrow CH_2{=}CHCH_2^\cdot + O_2 \ \Delta H = 12 \, kcal$$
$$CH_3CH_2O^\cdot \rightarrow CH_3^\cdot + CH_2O \qquad \Delta H = 13 \, kcal$$

The kinetic aspects of such reactions will be considered in some detail below.

Radical decompositions can also occur via a displacement reaction: this has frequently been proposed as the source of oxygen heterocycles in combustion reactions, for example,[24]

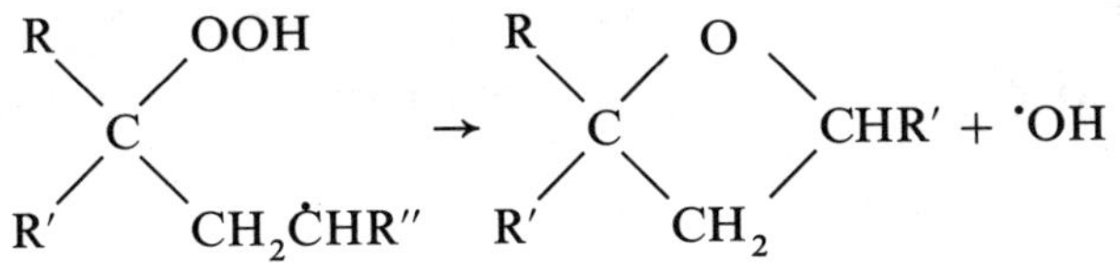

or

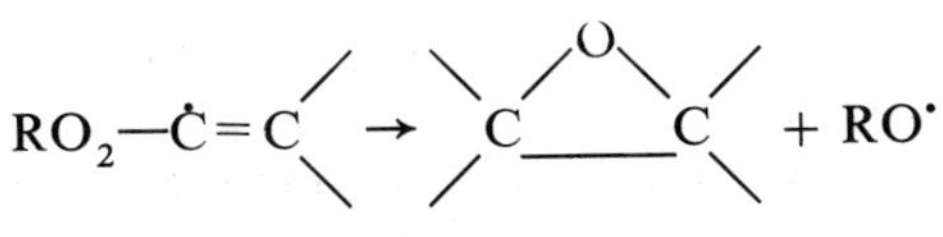

One final and rather unusual example of a unimolecular reaction is the intramolecular electron exchange which has been detected in the para-cyclophane radical anions $(3, n$ or $n_1 < 3)$ and the radical anion $(4, X = S$ or $O)$,

(III)   (IV)

If electron transfer is slow the esr spectrum is that expected from the localiza-tion of the free electron on one ring but when the exchange frequency is high the electron appears to be delocalized over the whole molecule and the e.s.r. spectrum exhibits hyperfine splittings due to all the magnetic nuclei of the whole molecule.[34, 35] Although the e.s.r. spectra of the inter-mediate cases are rather complex it has been possible by computer analysis of their main features, to determine the exchange frequency.

B. Bimolecular Reactions

(i and ii) Radical Combination and Disproportionation
These reactions usually occur competitively, for example,

$$C_2H_5{}^{\boldsymbol{\cdot}} + C_2H_5{}^{\boldsymbol{\cdot}} \xrightarrow{\ k_c\ } C_4H_{10} \qquad\qquad \Delta H = -82 \text{ kcal}$$

$$\xrightarrow{\ k_d\ } C_2H_6 + C_2H_4 \quad \Delta H = -60 \text{ kcal}$$

For alkyl radicals the ratio k_d/k_c ($=\Delta$) increases with increase in the number of hydrogen atoms which can be transferred in the disproportionation process. Studies with isotopically labelled ethyl radicals have shown that disproportionation occurs by a "head-to-tail" process,[36, 37]

$$CH_3CH_2{}^{\boldsymbol{\cdot}} + \overset{\displaystyle CH_2CH_2{}^{\boldsymbol{\cdot}}}{\underset{\displaystyle H}{\diagup}} \to CH_3CH_3 + CH_2{=}CH_2$$

Δ for alkoxy radicals is greater than that found for alkyl radicals and appears to be more nearly directly proportional to the number of removeable hydrogen atoms (see Chapter 4).

In the case of mixed radicals we can express cross-combination ratios as follows,

$$A + B \xrightarrow{k_{ab}} AB: \quad 2A \xrightarrow{k_{aa}} A_2: \quad 2B \xrightarrow{k_{bb}} B_2$$

If these reactions are similar then $k_{ab}/(k_{aa}k_{bb})^{\frac{1}{2}}$ should be 2, as has been reported for a number of radical pairs such as,[38]

A	B	$k_{ab}/(k_{aa}k_{bb})^{\frac{1}{2}}$
$CH_3^{\cdot}$	$C_2H_5^{\cdot}$	2·02
$CH_3^{\cdot}$	$n\text{-}C_3H_7^{\cdot}$	2·04
$CH_3^{\cdot}$	$CH_3COCH_2^{\cdot}$	1
$C_2H_5^{\cdot}$	$n\text{-}C_3H_7^{\cdot}$	1·93
$C_2H_5^{\cdot}$	$CCl_3^{\cdot}$	2
$C_2H_4Cl^{\cdot}$	$CCl_3^{\cdot}$	2·5

Disproportionations need not involve α-hydrogen atoms, for example[39, 40, 41]

$$2 Aro^{\cdot} \rightleftharpoons Aro^- + Aro^+$$

That combination reactions may also be complex has been seen already in the dimerization of trityl radicals,[42]

$$2 \, (C_6H_5)_3C^{\textbf{·}} \; \rightleftharpoons$$

(iii) *Transfer Reactions*

$$R^{\textbf{·}} + R'X \rightarrow RX + R''^{\textbf{·}} \quad \Delta H = D(R{-}X) - D(R'{-}X)$$

These are perhaps the most commonly studied of all radical reactions, particularly when X = H. Halogen atoms can also be transferred but the mechanisms and kinetics of such processes are complex. Where the radical R'' is sufficiently reactive the process can be repeated to give a chain reaction as in the reaction between hydrogen and chlorine,

$$Cl^{\textbf{·}} + H_2 \rightarrow HCl + H^{\textbf{·}}$$
$$Cl_2 + H^{\textbf{·}} \rightarrow HCl + Cl^{\textbf{·}}$$

But where R'' is so unreactive as not significantly to abstract an X atom in its turn and disappears by radical–radical reactions, the abstraction step is effectively isolated; one RX molecule is formed from each R· radical produced and the formation of RX is used to monitor the abstraction process. This is the principle of the radical trapping technique already mentioned. Typical examples of unreactive radicals R'' are iodine atoms and the resonance stabilized radicals, benzyl, allyl, phenoxyl and amino.

(iv) *Addition Reactions*
The essential feature of such processes is the addition of a radical to the π-system of an unsaturated molecule to give a larger radical which can combine, disproportionate or transfer as described in sections (i–iii) above to give small molecules or, alternatively, add to a further reactant molecule,

$$R^{\textbf{·}} + X{=}Y \rightarrow RXY^{\textbf{·}}$$

to give a larger radical which, if it in turn adds to a molecule of the reactant, ultimately produces a polymer. All of these various possibilities are nicely exemplified in the processes involved in the addition of CCl_4 to propylene:

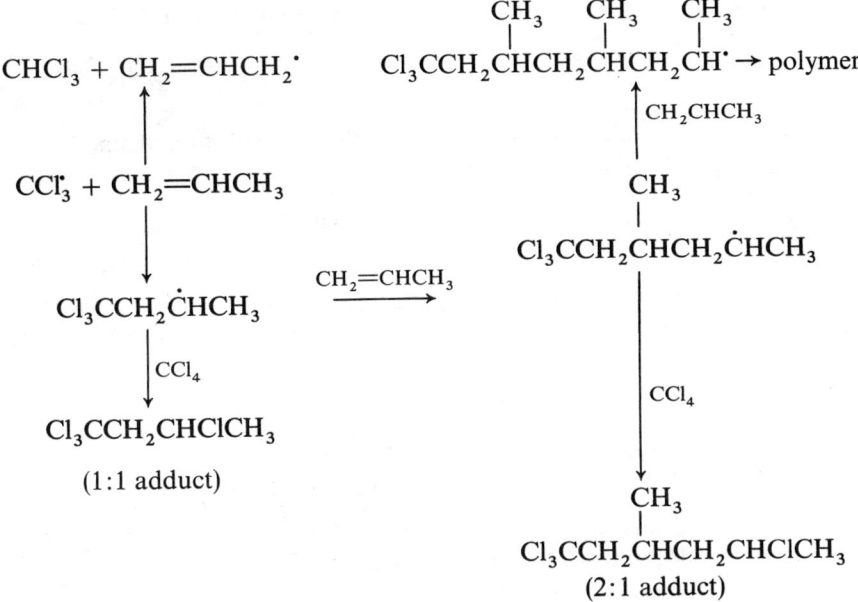

In general terms the production of polymers depends upon the addition reaction,

$$R^\cdot + \begin{smallmatrix}C\\ \diagdown\\ \diagup\\ D\end{smallmatrix}A{=}B \rightarrow R\overset{\cdot}{A}B\begin{smallmatrix}C\\ \diagdown\\ \\ \diagup\\ D\end{smallmatrix}$$

being favoured over transfer,

$$R^\cdot + \begin{smallmatrix}C\\ \diagdown\\ \diagup\\ D\end{smallmatrix}A{=}B \rightarrow RC + D\overset{\cdot}{A}{=}B$$

The A-factor for transfer is usually greater than that for addition (see Ch. 4) but the dominant factor is the difference in energy between that of the bond being broken, $D(C{-}A)$ and that being formed, $D(R{-}C)$. Where this is large, the former reaction will be favoured and the overall process will tend to produce polymers; where it is small, small molecules will be produced. Thus, molecules in which $D(R{-}C)$ is small, for example where $R^\cdot$ is resonance stabilized will tend to polymerize in radical reactions: for example, radical addition to styrene produces the resonance stabilized benzyl radical which, rather than abstract a hydrogen or other atom, will tend to add to a further molecule of styrene in a polymerization process,

$$C_6H_5CH{=}CH_2 + X^{\cdot} \rightarrow C_6H_5\dot{C}HCH_2X \xrightarrow{\ C_6H_5CH=CH_2\ } C_6H_5CHCH_2X$$

$$\underset{\text{etc.}}{\overset{\displaystyle |}{CH_2\dot{C}HC_6H_5}}$$

Conversely, these characteristics make it rather difficult to add small molecules such as CCl_4 to styrene,

$$C_6H_5CH{=}CH_2 + CCl_4 \rightarrow \text{polymers}$$
$$\rightarrow C_6H_5CHClCH_2CCl_3$$

the former process predominating over the latter.[43]

The following are some examples of small molecule free radical addition to multiple bonds.[44]

(a) *Polyhalomethanes*

$$RCH{=}CH_2 + CCl_3{}^{\cdot} \rightarrow R\dot{C}HCH_2CCl_3$$
$$R\dot{C}HCH_2CCl_3 + CCl_4 \rightarrow RCHClCH_2CCl_3 + CCl_3^{\cdot}$$

the ease of transfer decreasing in the order, $CX_3{-}X > CHX_2{-}X > CH_2X{-}X > CH_3{-}X$, X being the halogen atom.

(b) *Aldehydes and ketones*

$$R\dot{C}O + CH_2{=}CHR' \rightarrow RCOCH_2\dot{C}HR'$$
$$RCOCH_2\dot{C}HR' + RCHO \rightarrow RCOCH_2CH_2R' + R\dot{C}O$$

(c) *Amines*

$$R\dot{C}HNHR' + CH_2{=}CHR'' \rightarrow R\overset{\overset{\displaystyle NHR'}{\displaystyle |}}{C}HCH_2\dot{C}HR''$$

$$R\overset{\overset{\displaystyle NHR'}{\displaystyle |}}{C}HCH_2\dot{C}HR'' + RCH_2NHR' \rightarrow R\overset{\overset{\displaystyle NHR'}{\displaystyle |}}{C}HCH_2CH_2R''{}^{\cdot} + R\dot{C}HNHR'$$

(d) *Alcohols*

$$R\dot{C}HOH + CH_2{=}CHR' \rightarrow R\overset{\overset{\displaystyle OH}{\displaystyle |}}{C}HCH_2\dot{C}HR'$$

$$R\overset{\overset{\displaystyle OH}{\displaystyle |}}{C}HCH_2{}^{\cdot}CHR' + RCH_2OH \rightarrow R\overset{\overset{\displaystyle OH}{\displaystyle |}}{C}HCH_2CH_2R' + R\dot{C}HOH$$

(e) *Esters*

$$\dot{C}H(CO_2C_2H_5)_2 + CH_2{=}CHR' \rightarrow (CO_2C_2H_5)_2CHCH_2\dot{C}HR'$$
$$(CO_2C_2H_5)_2CHCH_2\dot{C}HR' + CH_2(CO_2C_2H_5)_2 \rightarrow$$
$$\rightarrow (CO_2C_2H_5)_2CHCH_2CH_2R' + \dot{C}H(CO_2C_2H_5)_2$$

A general feature of addition reactions is that the energy lost with the disappearance of the π-bond is about $50\,kcal\,mol^{-1}$. There are of course other consequent changes in the process each involving an energy element but usually the addition reaction is exothermic. As we shall see below this is a useful way of generating radicals containing excess energy over the equilibrium.

(v) Electron Transfer
This occurs between aromatic molecules and ions or between aromatic ion radicals and neutral molecules, e.g.

$$\text{naphthalene} + \text{naphthalene}^{-\bullet} \rightleftharpoons \text{naphthalene}^{-\bullet} + \text{naphthalene}^{45}$$

$$\text{stilbene} + \text{stilbene}^{-\bullet} \rightleftharpoons \text{stilbene}^{-\bullet} + \text{stilbene}^{46}$$

$$(pNO_2C_6H_4)_3C^- + (pNO_2C_6H_4)_3C^\bullet \rightleftharpoons (pNO_2C_6H_4)_3C^\bullet +$$
$$(pNO_2C_6H_4)_3C^{-\ 47}$$

$$(C_6H_5)_3C^+ + (C_6H_5)_3C^\bullet \rightleftharpoons (C_6H_5)_3C^\bullet + (C_6H_5)_3C^{+\ 48}$$

If the frequencies of these processes are close to those of the linewidths of the e.s.r. spectra of the radicals, they will cause a broadening of the lines because of the reduction of the lifetime of the electron in one state. Measurement of this broadening effect can give information about the kinetics of the exchange process. The rates of the above reactions fall in the range 10^7 to $10^9\,l.\,mol^{-1}\,s^{-1}$.

(vi) Aromatic Substitution
The overall reaction is

$$R^\bullet + ArY \rightarrow ArR + Y^\bullet$$

and most experimental work has been done with Y as a hydrogen atom and phenyl as the attacking radical. The mechanism involves[49] the formation of an intermediate resonance stabilized cyclohexadienyl radical which can undergo reactions of transfer, combination or disproportionation, (Scheme 1).

Phenyl substitution of substituted benzenes is not nearly as selective as in electrophilic or nucleophilic substitution where polarity effects are pronounced although no doubt if the substituting radical were also an ion the picture would become more complex. Steric effects also occur, for example in the reaction of phenyl radicals with t-butylbenzene, but in general the o-positions are attacked more readily than would be expected from purely statistical considerations. Also all substituents whether electron withdrawing or donating activate the nucleus to radical attack as predicted in the classification devised by Walter,[50] the cyclohexadienyl radical is stabilized both by electrophilic and nucleophilic substituents (class S)

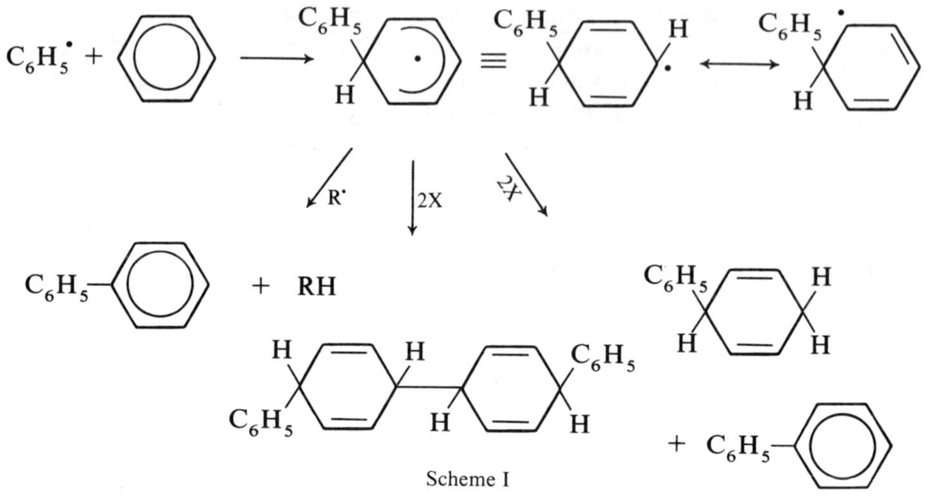

Scheme I

because an unpaired electron only and not an electron pair can be placed in the cyclohexadiene ring.

When the attacking phenyl radical is substituted the inductive effects of the substituent increase or decrease the electrophilicity of the radical in a manner which depends upon its nature. We can be sure that the effect is largely inductive because the site of reactivity of the phenyl radical, the half-filled sp^2 hybrid orbital is part of the σ and not of the π part of the molecule. Phenyl is thus a member of the class of σ-radicals which will be discussed in more detail below. The rates of aromatic substitution depend on the mutual affinity or disaffinity of the radical and substrate but the pattern of the substitution does not essentially change.

A considerable amount of work has been carried out on methyl substitution of aromatic compounds without, however, the wealth of analytical data available from arylation reactions. The method often used is competitive,

$$(CH_3COO)_2 \rightarrow 2CH_3{}^\cdot + CO_2$$
$$CH_3{}^\cdot + SH \rightarrow CH_4 + S{}^\cdot$$
$$CH_3{}^\cdot + M \rightarrow CH_3M{}^\cdot$$

where SH is a hydrocarbon with an easily abstracted hydrogen atom, for example iso-octane. Analysis for CH_4 with different unsaturated compounds M gives a "league table" of methyl affinities, or tendencies of the unsaturated substrates to undergo methyl substitution. These methyl affinities have been shown[51] to be related to localization energies, the differences in electron energy between, where one atom and two electrons have effectively been taken out of the π-system.

Streitwieser[12] has also suggested that in the case of methyl radical substitution instead of a cyclohexadienyl model a benzyl model would fit if the methyl is treated as an extension of the π-system, i.e.

The same author also claims that the use of this π-electron extension model can account for the substitution effects in arylation reactions, the effects of the substitution being explained using charge separated resonance structures.

(vii) Displacement Reactions

Discussion of these will be confined solely to reactions between radicals and saturated molecules although unfortunately the term has been applied to reactions which are more properly designated substitution reactions where a radical adds to a multiple bond. This is true of the first reported example,[52]

$$CH_3{}^{\cdot} + CH_3COCOCH_3 \rightarrow CH_3\overset{\overset{\displaystyle O^{\cdot}}{|}}{\underset{\underset{\displaystyle CH_3}{|}}{C}}COCH_3 \rightarrow CH_3COCH_3 + CH_3\dot{C}O$$

in which the second stage is driven by the energy of addition. Up to the present the author is not aware of a case of displacement at a saturated carbon atom but it does occur where atoms have available orbitals with lone pairs of electrons,[53] for example in chain transfer using disulphides,

$$P^{\cdot} + RSSR \rightarrow P{-}SR + RS^{\cdot}$$

in peroxide chemistry,

$$R^{\cdot} + R'OOR'' \rightarrow ROR' + {}^{\cdot}OR''$$

and with some silicon compounds,

$$CH_3^{\cdot} + SiCl_4 \rightarrow CH_3SiCl_3 + Cl^{\cdot}$$

(viii) *Non-radical Reactions*

The first of such reactions, not involving the unpaired electron function was the chlorination of porphrexide,[54]

but many more such reactions of nitroxides have now been reported.[1]

FACTORS AFFECTING THE REACTIVITY OF FREE RADICALS

The most important single factor controlling the reactivity of a free radical is its unpaired electron density; the closer this approaches to unity the greater will be the energy decrease when the unpaired electron becomes half of an electron pair and the greater will be the reactivity of the radical. Two factors act to oppose this, first the effects of conjugation which, in smearing out the unpaired electron density, reduce the energy to be gained in forming an electron pair and second, steric effects which tend physically to prevent close approach of the radical to a second chemical species, radical or molecule. We shall look at these effects in turn.

A. Resonance Effects

The larger the number of canonical forms in which the structure of the radical can formally be written, the greater its stability, for example the lack of reactivity of benzyl which, as we have noted, can be used in radical trapping experiments, is ascribed to electron delocalization shown by the structures,

Delocalization in diphenylpicrylhydrazyl which exists as a radical in the crystalline state is even more extensive,

$$Ph_2-\overset{\cdot\cdot}{N}-\overset{\cdot}{N}-Pic \longleftrightarrow Ph_2\overset{+\cdot}{N}-\overset{-}{N}-Pic \longleftrightarrow H-\underset{\cdot}{\bigcirc}=N-\overset{-}{N}-Pic$$

where pic = picryl. The stability of nitroxides can also be ascribed to resonance between structures of the sort,

$$\underset{R'}{\overset{R}{\diagdown}}N-O\overset{\cdot}{\longleftrightarrow} \underset{R'}{\overset{R}{\diagdown}}N^{+\cdot}-O^-$$

as indicated by e.s.r. spectroscopy which shows interaction between the unpaired electron and hydrogen atoms attached to the groups R. The mechanism of such interaction is readily seen from the canonical structures of diphenylnitroxide,

Nitroxides can also be looked upon as possessing a stable arrangement of five bonding electrons around the N—O group forming a three electron bond,[55]

$$
\begin{array}{cc}
\underset{R}{\overset{R}{\diagdown}}\!N \cdots O & \underset{R}{\overset{R}{\diagdown}}\!N\overset{X}{\underset{}{\underline{}}}\overset{X}{\underset{}{O}}\!-
\end{array}
$$

and this structure is compatible with the N—O bond length and vibration frequency (c. 1350 cm^{-1}) both of which are intermediate between those of nitromethane (i.r. vibration frequency 1564 cm^{-1}) and amine oxides (950–970 cm^{-1}).[56] The stability is explained in terms of the Linnett double quartet hypothesis[55] in which the six electrons of one spin and five of the other are arranged around the nitrogen and oxygen atoms in such a way that each has an octet; inter-electron repulsion is reduced to a minimum and there are five electrons involved in the bonding. Dimerization or the addition of another atom to the oxygen atom would not increase the number of bonding electrons but would increase inter-electron repulsion and is thus not energetically attractive.

Replacement of the oxygen atom by the less electronegative sulphur atom reduces the importance of structures of the sort,[57]

$$
\underset{}{\overset{\diagdown}{\diagup}}N^{+\,\cdot}\!-\!S^{-} \qquad \text{compared with} \qquad \underset{}{\overset{\diagdown}{\diagup}}N\!-\!S^{\cdot}
$$

and such thiyl radicals dimerize at the sulphur atom.

The reactivity of halogen substituted methyl radicals has also been explained on the basis of the electronegativity of the substituent atom. The i.r. spectrum of CBr$_3{}^{\cdot}$ [58] shows the force constant k_{C-Br} to be greater than that in C$_2$Br$_4$ and the explanation given involves resonance between structures of the sort,

$$
\underset{Br\ \ Br\ \ Br}{\overset{\dot{C}}{\diagup | \diagdown}} \quad \leftrightarrow \quad \underset{Br\ \ Br\ \ Br}{\overset{C^{-}}{\diagup | \diagdown}}
$$

Chlorine, being more electronegative is less likely to be involved in stabilization of this sort and the force constant k_{C-Cl} in CCl$_3{}^{\cdot}$ is less than that in C$_2$Cl$_4$ but greater than in CCl$_4$.

In the case of sp^2 hybridized halogen substituted methyl radicals it is possible to visualize back donation of electrons from the lone pair p_z orbital on the halogen atom to the half-filled p_z orbital on the carbon atom bearing the free electron where these orbitals are sufficiently close in energy for overlap to be effective, for example where X is the fluorine atom,

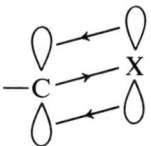

The electronegativity of fluorine is high and it would be expected that the σ-bond would be highly polarized towards the fluorine atom. Accompanying this charge transfer, however, the electronegativity of the carbon atom increases and that of the fluorine decreases and there could well be a partially compensating back donation from the full fluorine p_z orbital into the half-full carbon p_z orbital. The effect would be to increase the nucleophilicity of the carbon atom in a partial structure such as,

$$=C\begin{smallmatrix} \diagup F \\ \diagdown F \end{smallmatrix}$$

and indeed it has been noted[59] that hydrogen atoms, when used to attack such centres, appear to act as if they were electrophilic, no doubt being repelled by the increased negative charge. See also the reaction between methyl radicals and trifluorethylene.[62]

Aromatic free radicals in which the unpaired free electron is delocalized are stabilized on substitution both by electron accepting and electron donating groups. Both increase the extent of delocalization as can be seen by structures of the sort,

$$Ph_2\overset{\bullet}{C}\!\!\!\diagdown\!\!\!\!\bigcirc\!\!\!\!\diagup\!\!\!\overset{+}{N}\overset{\diagup O}{\diagdown O^-} \longleftrightarrow Ph_2C\!\!=\!\!\diamondsuit\!\!=\!\!\overset{+}{N}\overset{\diagup O^-}{\diagdown O^{\bullet}}$$

$$Ph_2\overset{\bullet}{C}\!\!\!-\!\!\!\bigcirc\!\!\!-\!\!OCH_3 \longleftrightarrow Ph_2C\!\!=\!\!\diamondsuit\!\!\overset{\bullet}{\diagdown}\!\!OCH_3 \longleftrightarrow Ph_2C\!\!=\!\!\diamondsuit\!\!=\!\!\overset{+\,\bullet}{O}CH_3$$

and both therefore displace the equilibrium,

$$Ar_3C\!-\!CAr_3 \rightleftharpoons 2Ar_3C^{\bullet}$$

to the right, i.e. in the same direction so that these radicals are classified as type S along with $Ar_3N^{+\bullet}$, $Ar_3B^{-\bullet}$ and $Ar_2N - N^{+\bullet}Ar_2$.[50] (But for a criticism of the classification of the radical $Ar_3N^{+\bullet}$ see Latta and Taft.[60])

Where the unpaired electron is associated with an atom which has more electrons than carbon there is also the possibility of delocalization of an unshared pair of electrons. Consider the imino radical,

$$Ar-N-Ar \qquad Ar-N-Ar$$

In the unsubstituted radical the energy of the 2p orbital, being higher than that of the sp^2 hybrid, the unshared pair of electrons occupies the latter orbital; the unpaired electron in the 2p orbital which overlaps with the π-electron systems of the aromatic rings is thus delocalized. But if electron withdrawing groups are put into the o- or p-positions the situation can become reversed and the lower energy state become that in which the unshared pair of electrons occupies the $2p_z$ orbital and is delocalized. The unpaired electron is then concentrated in the sp^2 orbital, the free electron density at the nitrogen atom is high, the radical reactivity also high and the equilibrium,

$$Ar_2NNAr_2 \rightleftharpoons 2(Ar_2N^{\cdot})$$

shifted to the left. Conversely, electron donating groups shift the equilibrium to the right and because of this opposite effect of electron withdrawing and donating substituents these radicals are placed in class 0 along with $Ar_2N\dot{N}Pic$, $Ar_2N\dot{N}COAr$, $ArNO_2^{\cdot}$, $Ar_2NO^{\cdot}$, $ArCO^{-\cdot}$ and (probably) $ArO^{\cdot}$.

These substituent effects are probably most complex in the hydrazyls which Walter writes as a resonance between the two structures,

$$Ar_2N\dot{N}Ar' \leftrightarrows Ar_2\overset{+}{N^{\cdot}}=\overset{-}{N}Ar'$$

$$(V) \qquad\qquad (VI)$$

Electron withdrawing substituents on Ar' and electron donating substituents on Ar will stabilize (VI) relative to (V) and shift the equilibrium

$$(Ar_2N-NAr')_2 \rightleftharpoons 2Ar_2\dot{N}-\dot{N}Ar'$$

to the right by decreasing the unpaired electron density on the nitrogen at which dimerization occurs.

B. Steric Effects

Although a carbon atom in a radical may carry a high unpaired electron density, the radical may still be unreactive if it has a geometry such that the site of the unpaired electron is inaccessible. This effect is shown at its best in the triphenylmethyl series in which the electron is extensively delocalized throughout the π-system of the phenyl rings but the maximum degree of delocalization is unattainable since the o-hydrogens of the aryl rings interfere with each other physically and constrain the radical to adopt a propeller-shaped configuration with an angle between the rings of about 30°.

The gas phase dissociation energy of hexaphenylethane has been estimated to be about 15 kcal mol^{-1}, in other words about 73 kcal mol^{-1} less than the unsubstituted ethane molecule. There would appear to be two possible reasons for this observation, first that the triphenyl methyl radicals are stabilized by delocalization and second that the phenyl rings surrounding the central carbon bearing the unpaired electron physically obstruct the dimerization process. This second factor, the steric effect, can be estimated by measuring the difference between the heats of hydrogenation of ethane and hexaphenylethane, thus,

$$CH_3CH_3 + H_2 \rightarrow 2CH_4 \qquad \Delta H = -13 \text{ kcal mol}^{-1}$$

$$(C_6H_5)_3CC(C_6H_5)_3 + H_2 \rightarrow 2(C_6H_5)_3CH \qquad \Delta H = -35 \text{ kcal mol}^{-1}$$

These reactions involve no change in the conjugation energy so that we can ascribe about 22 kcal mol^{-1} to the steric effect. The resonance effect for each triphenylmethyl radical compared to methyl is thus $\frac{1}{2}(73 - 22) = 25$ kcal mol^{-1}, a rather higher estimate than that of Benson[61] who calculated the resonance energy, at 16 kcal mol^{-1}, to be similar to that of benzyl in spite of the fact that esr studies have shown the unpaired electron density on the central carbon atom of triphenylmethyl to be 0·6 compared to 0·7 for benzyl.

Increasing the size of the o-groups in the hexaphenylethane series increases their tendency to dissociate from about 12% in the case of the unsubstituted compound to 87% with hexa-o-tolylethane although the spin density on the central carbon atom will increase as the radicals deviate more and more from coplanarity.

Steric effects are always very difficult to treat in any discussion of radical reactivity since there is no simple way of assessing their magnitude. Particularly is this the case when discussing the recombination of radicals or the addition of radicals to multiple bonds when steric hindrance to freedom in the transition state can lead to a low A-factor and perhaps even an activation energy effect.

REFERENCES

1. Forrester, A. R., Hay, J. M. and Thomson, R. H. (1968). "Organic Chemistry of Stable Free Radicals", Academic Press, London.
2. Robertson, A. J. B. (1959). "Mass Spectrometry", Methuen, London.
3. Laidler, K. J. (1965). *In* "Chemical Kinetics", McGraw Hill, New York.
4. Slater, N. B. (1959). "Theory of Unimolecular Reactions", Methuen, London.
5. Benson, S. W. (1960). "The Foundations of Chemical Kinetics", 234, McGraw Hill, New York.
6. Trotman-Dickenson, A. F. (1955). *In* "Gas Kinetics", Butterworths, London.
7. Schneider, F. W. and Rabinowitch, B. S. (1962). *J. Am. Chem. Soc.* **84**, 4215.
8. Setser, D. W. (1966). *J. Phys. Chem.* **70**, 826
9. Yee Quee, M. J. and Thynne, J. C. J. (1967). *Trans. Faraday Soc.* **63**, 2970.
10. East, R. L. and Phillips, L. (1967). *J. Chem. Soc. A* 1939.
11. Linnett, J. W. and Selley, N. J. (1963). *Z. Phys. Chem. (Frankfurt)* **37**, 402.
12. Streitwieser, A. Jr. (1961). *In* "Molecular Orbital Theory for Organic Chemists", Wiley, New York.
13. Hoare, D. E. and Pearson, G. S. (1964). *Adv. Photochem.* **3**, 83.
14. Rabinowitch, B. S. and Flowers, M. C. (1964). *Quart. Rev.* **18**, 122.
15. MacMillan J. and Price R. J. (1970). *J. Chem. Soc. B*, 337.
16. North, A. M. (1966). *Quart. Rev.* **20**, 421.
17. Isenberg, I. and Baird, S. L. (1962). *J. Am. Chem. Soc.* **84**, 3803.
18. Benson, S. W. (1961). *J. Chem. Phys.* **34**, 521. (*See also* Chapter 3.)
19. Phelan, N. F., Jaffe, H. H. and Orchin, M. (1967). *J. Chem. Ed.* **44**, 626.
20. Heller, C. A. and Gordon, A. S. (1958). *J. Phys. Chem.* **62**, 709.
21. Frey, H. M. and Walsh, R. (1969). *Chem. Rev.* **69**, 103.
22. Lossing, F. P., Ingold, K. U. and Tickner, A. W. (1953). *Disc. Faraday Soc.* **14**, 34.
23. Semenov, N. N. (1958). *In* "Some Problems of Chemical Kinetics and Reactivity", Pergamon Press, London.
24. Benson, S. W. (1965). *J. Am. Chem. Soc.* **87**, 972.
25. Endrenyi, L. and LeRoy, D. J. (1966). *J. Phys. Chem.* **70**, 4081.
26. Fish, A. (1964). *Quart. Rev.* 1964, **18**, 243.
27. Huyser, F. S. (1970). *In* "Free Radical Chain Reactions", 351, Wiley, New York.
28. Hickenbottom, W. J. (1936). *Nature* **142**, 830.
29. Beckwith, A. L. J. and Gara, W. B. (1969). *J. Am. Chem. Soc.* **91**, 5691.
30. Bennett, J. E. (1960). *Nature* **186**, 385.
31. Greene, F. D., Adam, W. and Knudsen, G. A., Jr. (1966). *J. Org. Chem.* **31**, 2087.
32. Wieland, H, (1911). *Ber.* **44**, 2550.
33. Hawley, D. M., Roberts, J. S., Ferguson, G. and Porte, A. L. (1967). *Chem. Comm.* 942.
34. Weissman, S. I. (1958). *J. Am. Chem. Soc.* **80**, 6462.
35. Harriman, J. E. and Maki, A. H. (1963). *J. Chem. Phys.* **39**, 778.
36. Wijnen, M. H. J., Steacie, E. W. R. (1951). *Canad. J. Chem.* **29**, 1092.
37. McNesby, J. R., Drew, C. M. and Gordon, A. S. (1955). *J. Phys. Chem.* **59**, 988.
38. Benson, S. W. and DeMore, W. B. (1965). *Ann. Rev. Phys. Chem.* **16**, 397.
39. Gomberg, M. and Cone, L. H. (1904). *Ber.* **37**, 3454; Schmidlin, J. and Garcia-Ban'a, A. (1912). *Ber.* **45**, 1344; Bowden, S. T. and Jones, W. J. (1928). *J. Chem. Soc.* 1149.
40. Marvel, C. S., Rieger, W. H. and Mueller, M. B. (1939). *J. Am. Chem. Soc.* **61**,

2769; Marvel, C. S., Mueller, M. B., Himel, C. M. and Kaplan, J. F. (1939). *J. Am. Chem. Soc.* **61**, 2771.

41. Calder, A. and Forrester, A. R. (1967). *Chem. Comm.*, 682
42. Lankamp, H., Nauta, W. Th. and MacLean, C. (1968). *Tet. Let.* 249.
43. Mayo, F. R. (1948). *J. Am. Chem. Soc.* **70**, 3689.
44. Pryor, W. A. (1966). "Free Radicals", McGraw Hill, New York.
45. Ward, R. L. and Weissman, S. I. (1957). *J. Am. Chem. Soc.* **79**, 2086.
46. Chang, R. and Johnson, C. S. Jr. (1967). *J. Chem. Phys.* **46**, 2314.
47. Jones, M. T. and Weissman, S. I. (1962). *J. Am. Chem. Soc.* **84**, 4269.
48. Lown, I. W. (1963). *Proc. Chem. Soc.* 283.
49. Williams, G. H. (1964). "Homolytic Aromatic Substitution", Pergamon Press, London.
50. Walter, R. J. (1966). *J. Am. Chem. Soc.* **88**, 1923.
51. Szwarc, M. and Binks, J. (1958). "Theoretical Organic Chemistry", 266, Butterworths, London.
52. Blacet, F. E. and Bell, W. E. (1953). *Disc. Faraday Soc.* **14**, 70.
53. Stirling, C. J. M. (1965). *In* "Radicals in Organic Chemistry", Oldbourne, London.
54. Piloty, O. and Schwerin, B. G. (1901). *Ber.* **34**, 1870, 2354.
55. Linnett, J. W. (1964). "The Electronic Structure of Molecules", Methuen, London.
56. Linnett, J. W. and Rosenberg, R. M. (1964). *Tetrahedron* **20**, 53.
57. Bennett, J. E., Sieper, H. and Tavs, P. (1967). *Tetrahedron* **23**, 1697.
58. Andrews, L. and Carver, T. G. (1968). *J. Chem. Phys.* **49**, 896.
59. Kilcoyne, J. P. and Jennings, K. R. Referred to in Moss, S. J. and Jennings, K. R. (1969). *Trans. Faraday Soc.* **65**, 415.
60. Latta, B. M. and Taft, R. W. (1967). *J. Am. Chem. Soc.* **89**, 5172.
61. Benson, S. W. (1965). *J. Chem. Ed.* **42**, 502.
62. Tedder, J. M., Walton J. C. and Winton, K. D. R. (1972). *Faraday Transaction I.* **68**, 1866.

CHAPTER 2
THE STRUCTURE OF RADICALS

In those favourable circumstances where free radicals can be prepared pure or at least in high concentration, a whole range of physical methods such as optical and nuclear magnetic resonance (n.m.r.) spectroscopy and X-ray or electron diffraction is at the disposal of the organic chemist. In addition, the magnetic properties vested in these species as a consequence of the presence of the unpaired electron can be exploited in the techniques of electron and double resonance spectroscopy.

In principle, and providing the response times of these physical methods could be made sufficiently short, they could be applied to the study of those transient free radicals produced at high concentration in the gas or condensed phase by flash photolysis but in practice only electron absorption spectroscopy has so far found much application in this field. Large numbers of free radicals have been detected both in the gas phase and in solution and the u.v. spectra and structures of the simpler species determined. The spectra of large radicals are more complex; phenyl has been analyzed partially and the spectra of large aromatic radicals assigned by systematic study of related series.[1] A promising advance in this field is the development of the giant pulsed laser.[2] A small number of radicals has also been observed in absorption and emission in flames and discharge tubes while species containing a few atoms only have been studied in discharge flow systems.[3]

The most powerful tool for the determination of the structure of free radicals of all sizes is, however, electron spin resonance spectroscopy (e.s.r.) and in view of the frequent references which are made in this book to results of e.s.r. analysis of radicals, it will be worth spending some time to survey briefly this method. More detailed information can be obtained in the publications listed in ref. 4.

E.S.R. SPECTROSCOPY

A free electron, as a spinning charged body has an associated magnetic moment, μ, given by,

$$\mu = -g\beta S \tag{1}$$

where g is the so-called spectroscopic splitting factor or more simply g-factor with a value close to 2 (actually 2·0023) for the free electron, β is the Bohr magneton and S the spin vector of the electron. When the electron is subjected to a magnetic field H_0 acting in the arbitrary reference direction, z, in space, the magnetic moment interacts with the field with energy, E, given by,

$$E = -\mu_z H_0 = g\beta M_s H_0 \tag{2}$$

where M_S is the magnetic spin quantum number with value of $+\frac{1}{2}$ or $-\frac{1}{2}$. The interaction gives rise to two energy states corresponding to whether the magnetic moment vector is parallel or antiparallel to the field, with a separation, in energy terms, of,

$$\Delta E = g\beta H_0 \tag{3}$$

or

$$h\nu = g\beta H_0 \tag{4}$$

For a frequency $\nu = 9\cdot5$ GHz (microwave frequency in the X-band region), the magnetic field is the conveniently attainable $H_0 = c.\ 3300$ gauss. The e.s.r. experiment consists in irradiating the sample held in the magnetic field with microwave radiation of a constant frequency while varying the magnitude of the field, H, until the conditions in eqn. (4) are satisfied; at this point energy is absorbed and detected electronically. This method is at present more dependable than varying the frequency of the incident radiation, the method adopted in optical spectroscopy. Suitable geometric arrangement of the microwave waveguides, for example into the shape of a "magic-T" allows the detection of the energy absorbed as an "out-of-balance" signal by means of a phase-sensitive detector. The output signal is normally displayed as the first derivative of the absorption line with the following general shape,

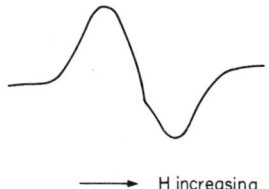

$\longrightarrow$ H increasing

The usefulness of e.s.r. spectroscopy derives from the interaction between the magnetic moment of the unpaired electron and that of any magnetic nuclei in the radical. In the simplest example, that of the hydrogen atom, the unpaired electron is located in a 1s orbital centred on the proton which has spin $I = \frac{1}{2}$. In an applied magnetic field, H_0, the magnetic moment vector of the proton is aligned parallel ($m_I = \frac{1}{2}$) or antiparallel ($m_I = -\frac{1}{2}$) to the field, the populations in the two levels differing by the small amount which is exploited in proton resonance spectroscopy but which is negligible in the e.s.r. context. The unpaired electron is thus subjected to two fields, $H_0 - H_1$ and $H_0 + H_1$ where H_1 is the magnitude of the field resulting from the magnetic moment of the proton. Two values of the applied field, $H = H_0 - H_1$ and $H = H_0 + H_1$ can satisfy eqn. (4) and the single absorption line is split into two, the area under and the intensity of each, which reflect the concentration of electrons subject to each field, being half that of the original signal. The selection rule for the transition of the electrons, $\Delta M_S = 1$ is that the nuclear spin should remain unaffected, i.e. $\Delta M_I = 0$. The spacing between the lines is known as the hyperfine splitting (h.f.s.) commonly denotaed by a and expressed in gauss or MHz (the conversion factor is 1 gauss = 2·803 MHz). For the hydrogen atom where the electron has complete 1s character the h.f.s., $a_H = 506$ gauss.

If the magnetic field of the electron interacts equally with that of two protons such as would occur in the radical moiety $-CH_2^{\cdot}$, then the signal from the single electron is split into three lines with intensity ratios of 1:2:1 and as the number of magnetically equivalent hydrogen atoms is increased to n the number of lines increases to $n + 1$ with intensity ratios given by Pascal's triangle,

$$
\begin{array}{ccccccccccc}
 & & & & & 1 & & & & & \\
 & & & & 1 & & 2 & & 1 & & \\
 & & & 1 & & 3 & & 3 & & 1 & \\
 & & 1 & & 4 & & 6 & & 4 & & 1 \\
 & 1 & & 5 & & 10 & & 10 & & 5 & & 1 \\
\end{array}
$$

etc.

Other nuclei with spin $= \frac{1}{2}$ are ^{13}C, ^{15}N, ^{19}F, ^{29}Si and ^{31}P. ^2D and ^{14}N have $I = 1$; ^{35}Cl, ^{37}Cl, ^7Li, ^{23}Na, ^{39}K and ^{33}S have $I = \frac{3}{2}$ and ^{17}O has $I = \frac{5}{2}$. Fortunately for the organic chemist, ^{12}C has no spin otherwise the e.s.r. spectra of organic radicals would become highly complex. The number of lines obtained by interaction of the unpaired electron with a nucleus of spin $= I$ is $(2I + 1)$.

As the number of magnetic nuclei contributing to the spectrum increases its pattern becomes more complex. Consider the situation where the unpaired electron is subject to a strong interaction with one nitrogen atom ($I = 1$, $a_N = 10$ gauss) and weaker interactions with two magnetically equivalent protons ($I = \frac{1}{2}$, $a_H = 1·5$ gauss). The spectrum observed will consist of three main lines of equal intensity, 10 gauss apart, each of which is further split into a 1:2:1 triplet, the lines of which are separated by 1·5 gauss (Fig. 1).

Some simple guide lines for the interpretation of e.s.r. spectra are given in ref. 5. Our interest in this book lies in the manner in which the e.s.r. spectrum of a radical may be interpreted so as to give information about the structure of the radical and it will be supposed that the spectrum has been fully analyzed.

FIG. 1.

The splitting of e.s.r. lines is of two kinds, isotropic and anisotropic; the first, occurring if the free electron has a finite probability of being at the position of the magnetic nucleus, for example a proton, that is its orbital includes some s-character, is independent of the orientation of the radical and is unaffected by its motion. The second on the other hand is governed by the spatial relationship between the electron and the magnetic nucleus. The hyperfine splittings of free radicals in the solid phase depend upon the orientation of the radical relative to the applied field but in a sufficiently mobile phase the tumbling and rotation of the radicals averages the anisotropic contribution to the h.f.s. to zero leaving the much simpler spectrum arising solely from isotropic effects. The reader is advised to consult a specialized e.s.r. work for further discussion of anisotropic effects; the following discussion deals only with isotropic splitting and that largely from hydrogen atoms.

If we represent the partial structure of aromatic radicals or simple π-alkyl radicals by,

$$>\overset{\displaystyle \cdot}{\underset{\displaystyle \cdot}{C}}-H$$

where the electron is in an orbital with a node at the position of the carbon atom, it is not obvious why the e.s.r. spectra of such radicals exhibit h.f.s. from the hydrogen atoms attached to the trigonal carbon. In other words, the orbital containing the free electron does not look to be able to embrace any proton s-character. The solution to this paradox is to be found in the effect known as spin polarization. The two possible arrangements of the electrons in the π orbital, the proton 1s orbital and the carbon sp^2 hybrid orbital which it overlaps (Fig. 2) are indistinguishable energetically on the simple Hückel molecular orbital theory but more detailed consideration shows that the unpaired electron affects the spin of the electron in the adjacent C_{sp^2} orbital in such a way that arrangement (ii) predominates to a slight extent over (i).

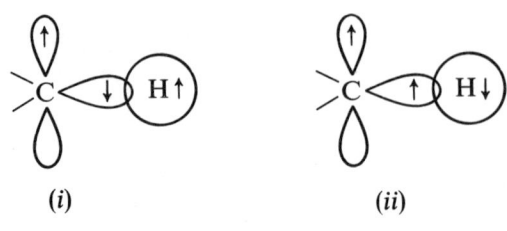

FIG. 2.

The result of this "configurational interaction" is a net unpaired negative spin density in the 1s orbital of the hydrogen atom, i.e. spin polarization and therefore hyperfine splitting. This h.f.s. has a negative sign but the sign is not important in the usual applications of e.s.r. The relation between the net unpaired spin density at the carbon atom (ρ) and the h.f.s. at the attached proton is given in an empirical manner by the McConnell equation,[6]

$$a_{\text{H}} = Q_{\text{H}}\rho \qquad (5)$$

where Q is a constant which varies from $-22{\cdot}5$ to about -30 gauss.

A similar empirical relationship links the h.f.s. of the hydrogen atoms of a freely rotating methyl group to the unpaired electron spin density in the p- or π orbital of an attached trigonal carbon atom,

$$a_{\text{H}} = Q_{\text{H}}^{\text{CH}_3}\rho \qquad (6)$$

$Q_{\text{H}}^{\text{CH}_3}$ appears to be constant at about $+27$ gauss for alternant hydrocarbons[7] (but see ref. 8). This interaction between the unpaired electron and the hydrogen atom of an attached methyl or other rotating alkyl group is almost certainly the result of a hyperconjugation mechanism[9]; the three 1s orbitals of the methyl hydrogen atoms can be combined in a formal mathematical manner to give one orbital with σ-symmetry and two with π-symmetry, one of which can overlap with the half-filled p orbital on the adjacent carbon atom. The way is then open for the electron to interact with the methyl protons and give hyperfine splitting with a positive sign. The mechanism has been confirmed using n.m.r. spectroscopy.[10] E.s.r. spectroscopy has thus been able to demonstrate that, although hyperconjugation may not be important in determining the properties of molecules it does occur in and presumably helps to stabilize radicals.

The McConnell equation[6] predicts that the overall width of the e.s.r. spectrum of an odd alternant hydrocarbon free mono-radical where there is no hyperconjugation should not exceed 30 gauss, but in the case of peri-naphthenyl (I),

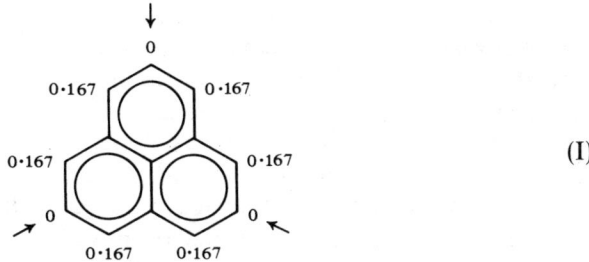

(I)

not only is the overall width about 49 gauss but the hydrogen atoms attached to the arrowed carbon atoms which, by simple Hückel theory should have zero spin density, contribute to the spectrum with $a_H = 2.2$ gauss.[11] Closer examination of the electron distribution throughout the radical shows that these positions have a negative spin density, i.e. the π-orbital on each of these carbon atoms bears an excess of β spins over α spins. The allyl radical, $CH_2 = CH—CH_2\,\dot{}$ also shows this effect, the central carbon atom having a net $\frac{1}{3}$ β spin and the two terminal atoms $\frac{2}{3}$ α spin giving the overall net position unpaired electron density of 1α.[12]

Equations similar to that of McConnell which have been developed for ^{13}C,[13] ^{14}N,[14] ^{19}F,[15] etc., are more complex because of the possible contribution to the splitting from structures in which there is an unpaired spin density in the 2s orbital. Exchange interaction in the fragment,

occurs between the unpaired electron in the π orbital on C_1 and the 1s and 2s electrons on C_1 giving a splitting proportional to ρ_1 and a smaller contribution results from spin polarization of the σ-bonds attaching C_1 to C_2 and C_3 by the unpaired electron in the π orbital of these atoms; this introduces negative spin density in the 2s orbital of C_1 proportional to ρ_2 and ρ_3 and the net splitting at the carbon atom is the sum,[14]

$$a_1 = Q_1\rho_1 + Q_2(\rho_2 + \rho_3) \tag{7}$$

where Q_1 is 30–35 gauss and Q_2 is about -14 gauss. a_1, thus being the difference between two fairly large quantities, is very sensitive to the distribution of the unpaired spin.

The e.s.r. spectrum of a radical can thus give information about its geometry and electronic arrangement. In favourable cases the sign of a_H can be determined by n.m.r. (see below) or by detailed e.s.r. studies of the radical orientated in the solid state; if the values found are negative and with a value larger (absolutely) than -30 gauss the radical is probably π and confirmation can be sought from measurements of a_{13_c}. If, on the other hand, a_H is very large as in the case of formyl, 137 gauss,[16] or the radical $\overset{\cdot}{C}NCH_2CONHCO\overset{\cdot}{N}H$, 80 gauss,[17] we can diagnose σ-character and similarly when $a_{\alpha\text{-}H}$ is very small or positive. The e.s.r. spectral characteristics of a number of σ-radicals are given in Table 1.

The reduction observed in $a_{\alpha\text{-}H}$ and the accompanying increase in a_{13C}, over the values expected for alkyl radicals, for radicals of the structure $R-\overset{\cdot}{C}H-OH$[18] could possibly be due to some s-character in the orbital associated with the unpaired electron, in other words deviation of the radical from the expected planar arrangement, but it could also reflect inductive effects and complications arising from back-donation from the full p orbital on the oxygen atom,[19]

$$H-O \overset{\text{}}{\longleftarrow} C \overset{H}{\underset{R}{\diagup}}$$

It has been suggested[20] that 16% of the unpaired spin is associated with the oxygen atom and 84% with the carbon. Increasing the number of oxygen substituents further acentuates these effects.[21]

Unfortunately the gas-phase-free-radical chemist is unlikely to benefit greatly from the detailed structural information, high sensitivity (detection down to about 10^{-10} molar) and great selectivity (detection only of paramagnetic species) of e.s.r. spectroscopy. In the gas phase, atoms and diatomic radicals are suitable subjects for study but as the radical becomes larger the electron spin angular momentum vector tends to couple with the radical's rotational angular momentum vector, in essence giving to each rotational "temperature" of the radical a separate e.s.r. spectrum. The rotational energy of a radical is spread over a large number of levels so that its e.s.r. spectrum becomes unresolvable and its intensity, which is in any case probably low for concentration reasons, is spread over many gauss so as to become indistinguishable from the base line of the spectrum.

N.M.R. SPECTROSCOPY[22]

This technique is also unsuitable for gas phase application but in the liquid phase and under certain conditions its use may reveal some of the detailed

Table I
E.s.r. spectra of some σ-radicals

Radical	a_H (gauss)
(cyclopropyl radical)	$a_{\alpha\text{-}H} = 6\cdot51^a$ $a_{\beta\text{-}H} = 23\cdot42$
(oxiranyl radical)	$a_{\alpha\text{-}H} = 2\cdot33^b$ $a_{\beta\text{-}H} = 0\cdot61$
$\begin{array}{c} H \\ H \end{array} C{=}C \begin{array}{c} \cdot \\ H \end{array}$ (vinyl radical)	$a_{\alpha\text{-}H} = 16\cdot0^c$ $a_{\beta\text{-}trans\text{-}H} = 68$ $a_{\beta\text{-}cis\text{-}H} = 34$
$\begin{array}{c} H \\ DOOC \end{array} C{=}C \begin{array}{c} H \\ \cdot \end{array}$	$a_{\alpha\text{-}H} = 13\cdot5^d$ $a_{\beta\text{-}H} = 58$
(phenyl radical)	$a_{o\text{-}H} = 18\cdot1^e$ $a_{m\text{-}H} = 6\cdot4$ $a_{p\text{-}H} = 3\cdot0$
$\begin{array}{c} H \\ \cdot \end{array} C{=}O$	$a_H = 137^f$
(benzoyl radical) CO	$a_{o\text{-}H} = 0\cdot1^g$ $a_{m\text{-}H} = 1\cdot16$ $a_{p\text{-}H} = 0\cdot1$
$CNCH_2CONHCO\dot{N}H$	$a_H = 80^h$

[a] Fessenden, R. W. and Schuler, R. H. (1963). *J. Chem. Phys.* **39**, 2147; [b] Dobbs, A. J., Gilbert, B. C. and Norman, R. O. C. (1971). *J. Chem.Soc. A*, 124; [c] Cochran, E. L., Adrian, F. J. and Bowers, V. A. (1964). *J. Chem. Phys.* **40**, 213; [d] Iwasaki, M. and Eda, B. (1970). *J. Chem. Phys.* **52**, 3837; [e] Bennett, J. E., Mile, B. and Thomas, A. (1966). *Proc. Roy. Soc. A*, **293**, 246; [f] Adrian, F. J., Cochran, E. L. and Bowers, V. A. (1962). *J. Chem. Phys.* **36**, 1661; [g] Krusic, P. J. and Rettig, T. A. (1970). *J. Amer. Chem. Soc.* **92**, 722; [h] Pau, P. W. and Lin, W. C. (1969). *J. Chem. Phys.* **51**, 5139.

characteristics of free radicals such as the signs of their coupling constants. The principle of n.m.r. is closely similar to that of e.s.r. and the width of the n.m.r. line, in frequency terms is given by,

$$\Delta v = \tfrac{1}{2}\pi\tau$$

where τ is the time during which the proton magnetic vector remains undisturbed; the longer this time the sharper will be the n.m.r. line. Since the electron has a magnetic moment some 1000 times that of the proton, the presence of paramagnetic material in an n.m.r. sample will tend to reduce the lifetime of a proton spin state and so broaden the n.m.r. line.[23] A further effect is a contact shift in the n.m.r. spectrum[24] moving the lines in a manner depending on the sign and magnitude of the electron-nuclear coupling constant, upfield for positive spin densities and downfield for negative. As explained in ref. 25 the n.m.r. spectral lines can be sharpened at high concentrations of free radicals giving, in favourable conditions, not only the signs of the coupling constants but also the magnitude of those that are too small to be resolved using a conventional e.s.r. spectrometer. Some coupling constants and their signs, which have been determined by this method are given in Table II.

INFRARED (I.R.) SPECTROSCOPY

The i.r. spectra of radicals trapped in solid matrices are also a fruitful source of radical structural information and a triumph of the technique was the elucidation of the structure of methyl.[26] Acyl radicals have also been examined[27] and the bent shape of formyl confirmed (HCO = 123° 8′); the force constant of the C—O bond turns out to be 13·7 mdyne Å^{-1} corresponding to a bond order of 2·3. The acetyl radical has also been shown to have a bent configuration and a similar C—O bond order,[28]

EMPIRICAL CONSIDERATIONS

As the first part of this chapter has shown the structure of a radical may be calculable from its optical or magnetic resonance spectrum provided that it can be prepared predictably and in a suitable physical form. A number of ingenious methods have been devised to this end but we are still a long

Table II

N.m.r. spectra of some radicals

R	$a_{t\text{-Bu-H}}$	a_N	a_{OCH_3-H}	$a_{\beta\text{-H}}$	$a_{\gamma\text{-H}}$	$a_{\delta\text{-H}}$
H[a]	+0·063	+1·78	+0·935	+2·32		
Δ (anti)	+0·068	+1·85	+0·82	−2·02		
□ (anti)	+0·068	+1·82	+0·70	−1·99	+0·104	−0·026
CH_3[b]	+0·072	+1·82	+0·83	−1·70		
CH_2CH_3	+0·065	+1·84	+0·81	−0·95		
$CH_2CH_2CH_3$	+0·065	+1·81	+0·81	−0·94	+0·128	
$CH_2CH(CH_3)_2$	+0·065	+1·86	+0·85	−1·02	+0·219	
$C(CH_3)_2$	+0·073	+1·86	+0·16		+0·134	

Solvent	a_N	$a_{\alpha\text{-H}(1)}$	$a_{\alpha\text{-H}(2)}$	$a_{\beta\text{-H}}$	$a_{\gamma(CH_3)\text{-H}}$	$a_{Ring\,CH_3\text{-H}}$
$CDCl_3$[c,d]		−2·93	−1·41	+0·18	−0·038	−0·21
Toluene[e]	7·25	2·96	1·22			0·22

[a] Kreilick, R. W. (1968). *J. Amer. Chem. Soc.* **90**, 5991; [b] Yamuchi, F. and Kreilick, R. W. (1969). *J. Amer. Chem. Soc.* **91**, 3429; [c] Kreilick, R. W., Becher, J. and Ullman, E. F. (1969). *J. Amer. Chem. Soc.* **91**, 5121; [d] N.m.r. spectrum; [e] e.s.r. spectrum.

way from a satisfactory theory of radical reactivity based on structure determinations. In the first place not all radicals have been detected and of those that have, the structure of a few only have been analyzed unambiguously and second, wherever possible, spectroscopic measurements are made on ground state species whereas during reaction radicals may contain excess energy over the equilibrium and it is possible to visualize situations in which such radicals may react in a configuration different from that of the ground state. We can illustrate such a possibility by considering the case of the dissociation of a substituted ethane, (1)

$$R_3C\text{—}CR_3 \rightarrow 2R_3C^{\cdot} \tag{1}$$

involving the rupture of the central C—C bond to give in the extreme case two substituted methyl radicals in which the trigonal carbon atom, i.e. that bearing the free electron, may be hybridized in a planar, sp^2, or tetrahedral, sp^3, arrangement, the unpaired electron being in a p or sp^3 orbital respectively or in other words the radical being π or σ. (Intermediate stages of hydridization with the HCH angle lying between $120°$ and $109\frac{1}{2}°$, can occur, as for example in the $\dot{C}HF_2$ radical,[29] but for simplicity the two extreme forms only will be discussed.)

Mechanistically, the reaction to give pyramidal radicals involves a transition state differing from the original molecule only in the length of the central bond,

$$\begin{array}{ccc} \overset{R}{\underset{R}{\overset{R}{\diagdown}}}C\text{—}C\overset{R}{\underset{R}{\diagup}} & \longrightarrow & \overset{R}{\underset{R}{\diagdown}}C\text{----}C\overset{R}{\underset{R}{\diagup}} & \longrightarrow & 2\ \overset{R}{\underset{R}{\diagdown}}C\!\!\diagup \end{array} \tag{1a}$$

The effect is as though the central bond has suddenly been removed and the reaction may be called an instantaneous bond dissociation with an associated instantaneous bond dissociation energy (i.b.d.e.).

On the other hand, dissociation to give planar radicals proceeds through a transition state where one can imagine the two halves of the molecule flattening,

$$\begin{array}{ccc} \overset{R}{\underset{R}{\overset{R}{\diagdown}}}C\text{—}C\overset{R}{\underset{R}{\diagup}} & \longrightarrow & C\text{—}C & \longrightarrow & 2\ R\text{—}C \end{array} \tag{1b}$$

Given that hydrocarbon radicals in their ground state are normally stable in a configuration planar around the carbon atom bearing the unpaired electron, reaction (1b) is the usual dissociation reaction and the energy required is the bond dissociation energy (b.d.e.) which is lower than the i.b.d.e.

Another effect arises from the loose transition state of reaction (1a) in which the two halves of the hypothetical species are freer to rotate and vibrate than in the original molecule; this reaction would therefore be expected to have a high entropy of activation, or in other words an A-factor greater than the "normal" value of about $10^{13}\ s^{-1}$. In (1b) the transition state involves a reduction in the s-character of the C—C bond; this may be accompanied by the pulling together of the two halves of the molecule and the concomitant flattening of the R_3C- groups making the transition state more crowded. Under such conditions internal rotation and vibration are hindered, the entropy of the transition state is less than that of the parent molecule and the reaction has a lower than normal A-factor. Looked at in the reverse sense one could say that the approach of two planar $R_3C^{\cdot}$ radicals preparatory to combination is limited to a narrower angle than would be the case for the recombination of two pyramidal radicals, reaction $(-1a)$.

Turning to the energy requirements of the two processes, it is reasonable to suppose that reaction (1b) has the lower activation energy since during its course the C—C bond is changing from $C_{sp^3}-C_{sp^3}$ to the much weaker C_p-C_p with associated strengthening of the C—R bonds which change from $C_{sp^3}-R$ to $C_{sp^2}-R$. In other words the energy released during the transition of the radical from a pyramidal structure to the ground state planar arrangement contributes to the driving force of the reaction. No such assistance is available in the case of the instantaneous bond dissociation (1a).

For dissociations leading to radicals which are planar in the ground state, then we can depict the potential energy diagram for the two dissociations schematically as in Fig. 3.

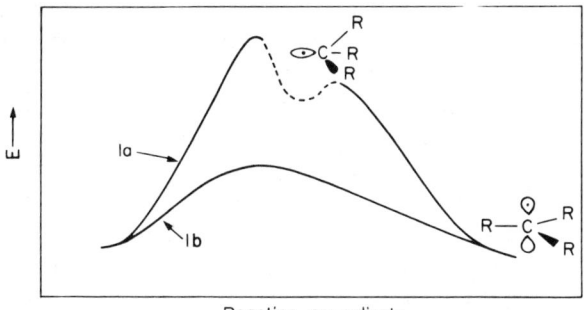

Reaction co-ordinate

FIG. 3.

From this simple picture we can see that the competition between (1a) and (1b) will depend on that between the increased A-factor of the former and the decreased activation energy of the latter. If sufficient energy is available so that the activation energy is not a rate determining consideration, then dissociations of this sort will tend to go by reaction (1a) (for example σ-t-butyl radicals, have been detected by e.s.r. in the products of γ-irradiated t-butyl chloride.[54]) but this is not usually the case under the sort of reaction conditions normally used in the laboratory. Also, if there are large steric factors associated with the formation of the transition state of path (1b) there might be a tendency to follow (1a). This may occur in the dissociation of hexaphenylethane where the A-factor, which is expected to be low because of the considerable stabilization energy of the triphenyl-methyl radicals, is in fact close to normal. (The reverse of this reaction which is accompanied by a fairly high activation energy of about 9 kcal mol^{-1}, may not be a simple recombination.[30])

It is the purpose of the remainder of this chapter to discuss the factors which control the tendency of free radicals to exist in the π or σ state. We shall start by considering the methyl radical and our first assumption will be that the carbon atom has been excited into its "valence" state by the promotion of an electron from the 2s orbital of the $1s^2 2s^2 2p^2$ ground state to the empty 2p orbital, giving $1s^2 2s 2p^3$ and the half-filled orbitals have hybridized to give, in this case of trigonal methyl,

$$^3P_{C\,s^2p^2} \xrightarrow{\; c.\ 96\ kcal \;} {}^5S_{C\,sp^3} \xrightarrow{\; c.\ 70\ kcal \;} \text{valence state hybrids.}$$

The valence state hybrids are not a spectroscopic state. This excitation energy probably makes the largest contribution to the overall energy of the radical but its value is in doubt. Even in the case of methane the picture is complicated since the bonding is not purely sp^3 hybrid; resonance (con-figurational interaction) with certain s^{2}p^2 hybrids reduces the total energy to a value which Coulson has estimated[31] to be approximately 100 kcal mol^{-1}. The even lower value of 39 kcal mol^{-1} has, however, recently been suggested.[32] We shall return to this question in Chapter 3.

Experimentally it has been shown that the methyl radical in its ground state is planar or very nearly so;[33, 34] no inversion was observed by these authors in the e.s.r.[33] or u.v. spectra[34] of the radical. On the other hand from the out-of-plane bending force constants of the methyl radical in a solid argon matrix (at 1383 cm^{-1}), the inversion energy of the methyl radical has been estimated to be only 4·2 kcal.[26] Recently it has also been shown that the variation of the h.f.s. for H and ^{13}C in the e.s.r. spectrum of the methyl radical over the temperature range 245–333°K can be satisfactorily explained, from empirical calculations on a pyramidal methyl model, as arising from increase in the out-of-plane bending of the radical with increase in temperature.[35] This detection of out-of-plane distortion of the methyl radical lends support to the suggestions which are discussed in Chapter 4, that we cannot be certain that in the gas phase at temperatures commonly used for, say, pyrolysis reactions or in the presence of the sort of energy involved in photolysis reactions, all methyl radicals present and reacting will be planar.

The factors controlling the configuration of a radical such as methyl where electronegativity effects are negligible are (*a*) the energy of the unpaired electron(s), (*b*) the energies of the bonds in the radical and (*c*) the H—H interaction. These can be depicted in the case of methyl by a simple "Walsh-type"[36] diagram of the energy changes which are supposed to accompany the inversion of the methyl radical (Fig. 4).

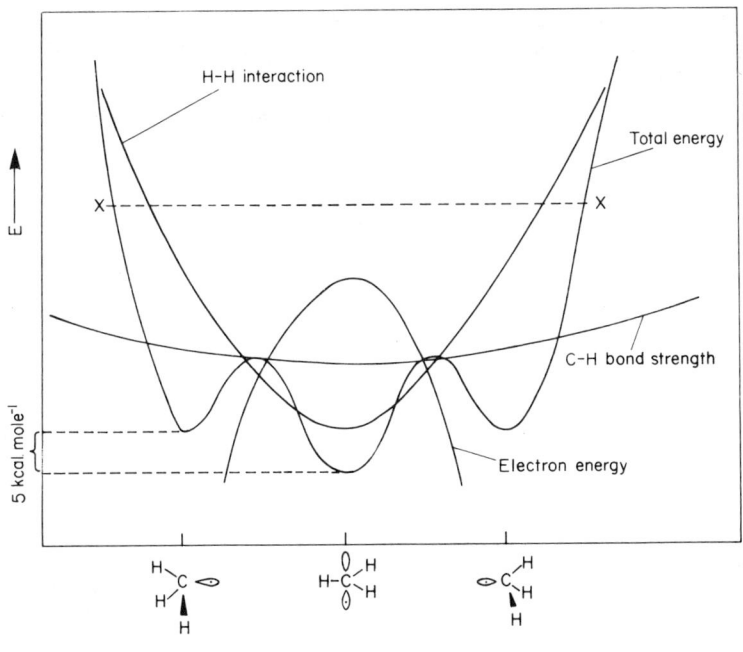

FIG. 4.

As the tetrahedral structure flattens out the binding energy of the unpaired electron decreases since it changes its orbital from sp^3 to p in which electrons are much less tightly held. At the same time there is an associated increase in the C—H bond energy and decrease in the H—H repulsion energy as the hybridization around the carbon atom changes from sp^3 to sp^2. The total energy is shown tentatively drawn as a double barrier to inversion. The energy barrier to inversion should be rather lower than that of ammonia (*c*. 6 kcal mol^{-1}) since the half-filled sp^3 orbital will experience less resistance to inversion than that carrying the lone pair of electrons in the ammonia molecule. The stabilization energy of the most stable configuration of the methyl radical (*b*) has been calculated to be about 5 kcal mol^{-1} (see Chapter 3 below) and Kibby and Weston have calculated pyramidal methyl to have some 10 kcal mol^{-1} of energy greater than the ground state.[37]

If we consider for a moment the methyl radical excited sufficiently to be inverting between the two energy barriers at a level above that of the maxima of the triple potential well, for example at X—X, it is clear that since the inversion slows down and reverses close to the walls of the potential well, the radical will spend on the average most of its time in a pyramidal or near pyramidal configuration. To other species in a system with such a methyl radical, it will appear to have mostly a σ-structure.

A similar diagram can be drawn for the vinyl radical but in this case the radical is known[38] to have an energy barrier to inversion of 1·4 kcal mol^{-1} and to have a minimum energy at an angle $\theta = 155°$ (see Fig. 5). The H—H interaction is ignored.

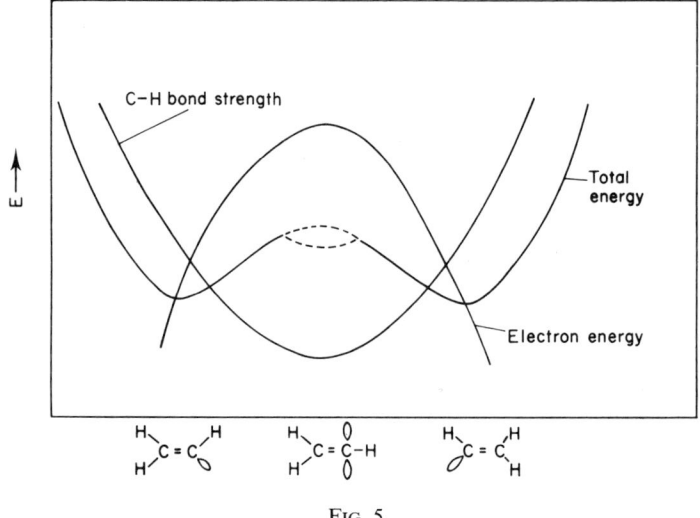

FIG. 5.

With increase in the number of carbon atoms and electrons it becomes more difficult to predict the ground state configuration of radicals but it is instructive to consider how this is affected by the various contributory factors mentioned above. The first factor, the valence state excitation energy needs no further discussion at this stage.

The second most important contribution to the energy of a carbon radical is that of the unpaired electron which can be reduced by overlap of the orbital containing the electron with other orbitals of the radical, most frequently the π-bonding system. A good example of this is the benzyl radical where the half-filled p_z orbital of the trigonal carbon atom which carries the electron overlaps with the π-bonding orbital of the benzene ring. The path length available to the unpaired electron is thus increased and it has a certain positive probability of being at any position of the ring as well as the trigonal carbon atom. This electron density at the trigonal carbon atom is found to be 0·7.[39] Pictorially this phenomenon of delocalization can be represented by writing the formula of the radical in a number of "canonical" forms, thus,

Increase in the path length further decreases the energy of the electron.

The delocalization energy also decreases ΔH_1; indeed ΔH_1 for unimolecular decompositions producing conjugated radicals has been used as a measure of delocalization energy. For example the increase in this quantity (ΔM) in going from benzyl to methyl is given by the equation,

$$\Delta M = D(CH_3-H) - D(C_6H_5CH_2-H)$$
$$= 104 - 85$$
$$= 19 \, \text{kcal mol}^{-1}$$

and for allyl,

$$\Delta M = D(CH_3-H) - D(CH_2=CH-CH_2-H)$$
$$= 104 - 85$$
$$= 19 \, \text{kcal mol}^{-1}.$$

These values compare reasonably well with those calculated from Hückel molecular orbital theory.[40]

As mentioned above, hyperconjugation in a radical also gives rise to delocalization; ethyl, for example can be written in a molecular orbital formation as follows,

The values of $a_H^{CH_3}$ and therefore hyperconjugative effects are reduced when the free rotation of the methyl group is hindered.[41]

In the case of nitrogen- or oxygen-containing radicals reduction of the energy of the unpaired electron by delocalization as for example in anilino or phenoxyl, also increases the stability of the radical but complications arise in the case of species with the partial structure.

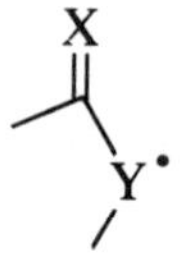

Here the unpaired electron is (formally) localized on an atom Y such as N or O which also has an associated unshared pair of electrons, the energy of which can also be reduced by delocalization in a π-system including an electronegative element X. In principle there are two possible structures, the one involving delocalization of the unpaired electron being a π-radical and the other, in which the orbital of the unshared pair of electrons overlaps with that of the π-system leaving the unpaired electron on Y, being a σ-radical. The succinimidyl radical (II) is a good example of this type of structure,[42]

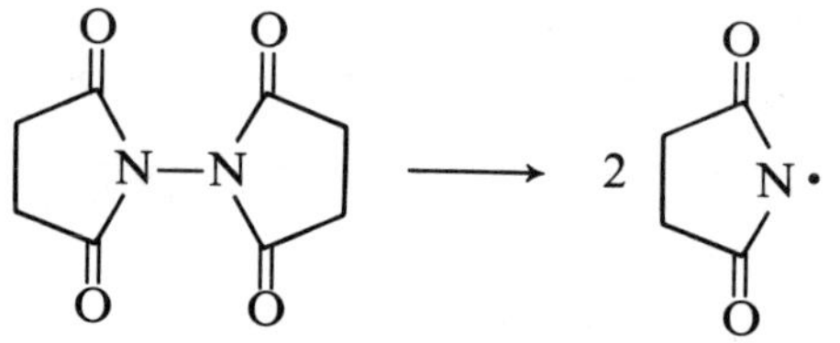

(II)

For simplicity we shall again assume that the nitrogen atom has been promoted to its valence state so that the parent molecule can be angular, sp^3, or planar, sp^2, around the nitrogen atom. Reduction of the energy of the unshared pair of electrons in the parent molecule by delocalization over the π-system containing the two electronegative oxygen atoms decreases the energy of the latter compared to that of the former arrangement and the neutral molecule is planar around the nitrogen atoms. The radical can in principle exist in two forms, one (a) a σ-radical with the unpaired electron in a sp^2 hybrid orbital and the unshared pair delocalized and the other (b) a π-radical where the unpaired electron is delocalized. The former dissociation, into σ radicals, is, as we have seen, associated with the higher bond dissociation energy,

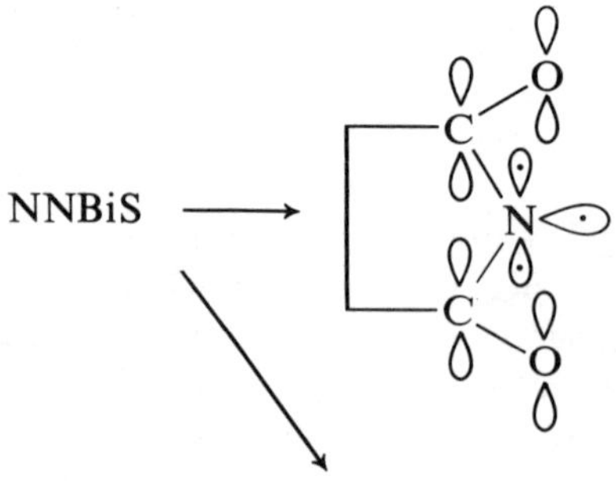

NNBiS $\longrightarrow$

(a)

(b)

The relevant orbitals in the radical are the five overlapping p-orbitals giving the π-system, the remaining orbitals on the nitrogen not involved in the σ bonding and the orbitals on the oxygen atoms containing the unshared pair of electrons:

$$— 5$$

$$— 4$$

$$\uparrow\downarrow 8 \;\uparrow\downarrow 7 \;\uparrow\downarrow 3 \;\uparrow 6$$

$$\uparrow\downarrow 2$$

$$\uparrow\downarrow 1$$

Whichever form is the more stable depends upon whether $e_6 <$ or $> e_3$; rough molecular orbital calculations suggest the latter and a σ-arrangement for the succinimidyl radical, a conclusion which is borne out by its chemistry (see p. 23). One difficulty attaching to predictions of this sort is that even if e_6 were less than e_3, if the radicals were initially produced in the σ-form in order for them to attain the more stable π-form, some mechanism would have to be available for the transition of the electron from ψ_3 and ψ_6 and, should this prove difficult, the radical might still exhibit σ-reactivity.

The important feature of this discussion is that it emphasizes that substitution of a group with a high electronegativity associated with a p or π-orbital and where the unpaired electron is located on an atom with an unshared pair of electrons in the L-shell, i.e. of the general structure

R—C=B where A and B are NH or O, will tend to favour the σ structure of the radical.[43] In this connection we have already noted Walter's classification of radicals into S and O[44] where in the one case only the unpaired electron is delocalized while in the other either the unpaired electron or an unshared pair of electrons can be delocalized.

The third factor controlling the structure of a free radical which has been mentioned briefly in the discussion of the "Walsh-type" diagrams is the strength of the σ-bonds around the atom bearing the free electron.

Since the ground state energy difference between σ- and π-radicals probably does not exceed 1 eV in normal cases, factors which tend to alter the strengths of these bonds and hence the energies of their electrons will be important in determining which of the two structures is the more stable. Let us look, for example at the radical $R_3C^{\cdot}$. If the structure of R remains unchanged while the hybridization around the carbon atom changes from sp^3 to sp^2 with concomitant change in the orbital containing the unpaired electron from sp^3 to p, the greater overlap potentialities of the sp^2 orbital strengthen the C—R bond and favour the π-form of the radical. The same sort of consideration applies to molecules and radicals containing heteroatoms. Comparing methyl with ammonia we find that in the latter case, two electrons in a p orbital have an energy which outweighs the extra strength of N_{sp^2}—H over that of N_{sp^3}—H giving ammonia a pyramidal structure. In water similarly the HOH bond angle, 104° 31′, is close to tetrahedral. In the example of methyl, the extra strength of the C_{sp^2}—H bonds predominates and the radical is planar with the angle HCH about 120°.

The fourth factor is the electronegativity of a group or atom attached to the trigonal carbon atom. Owing to the high binding energy of electrons in s orbitals the electronegativity of an atom or group attached to another increases with the percentage s-character of the orbital of the former which is involved in the linkage. This is clearly seen from Table III which also gives the calculated electronegativities of a number of atoms and groups.[45] The column headed a corresponds approximately to the Mulliken or Pauling scales while the b column is a measure of how the electronegativity of the atom or group changes with changes in the attached group; the greater the value of b the greater the electronegativity decrease with charge displacement. b is thus a rough measure of the charge which the atom or group can accommodate without much change in its electronegativity. Single atoms can tolerate less charge before change in their electronegativity than can groups of atoms.

To see how this is relevant to the structure of free radicals, consider the case of the radical fragment, $-CF_2$, which can exist in one of two extreme structures, the tetrahedral (i) and the flat planar (ii),

(i)   *(ii)*

In (i) the bonding orbital overlapping that of the electronegative fluorine atom is sp^3 and has thus a lower s-content than the corresponding sp^2

TABLE III

Group electronegativities[a]

Group	Hybridization	a	b
C	p	5·80	10·93
	sp^3	7·98	13·27
	sp^2	8·79	13·67
	sp	10·39	14·08
	s	14·96	12·10
CH_3	sp^3	7·37	3·24
CH_3CH_2	sp^3	7·40	1·85
CH_2F	sp^3	8·38	3·46
CHF_2	sp^3	9·55	3·73
CF_3	sp^3	10·90	4·03
CCl_3	sp^3	9·07	2·93
CBr_3	sp^3	8·32	2·53
C_6H_5	sp^2	8·03	1·21
$CH_2{=}CH$	sp^2	7·79	2·63
$HC{\equiv}C$	sp	9·25	4·55
H	s	7·17	12·85
F	p	12·18	17·36
Cl	p	9·38	11·30

[a] Huheey, J. E. (1964). *J. Phys. Chem.* **68**, 3073; (1965). **69**, 3284; (1966). **70**, 2086.

orbital of (ii). In the former structure therefore the electrons in the sp^3 orbitals are less firmly bound to the carbon atom and a greater degree of charge movement from C to F is possible than can occur in (ii). This effect thus counterbalances the normal tendency of the radical to be planar and helps stabilize the σ-structure. The greater the number of electronegative substituents, of course, the greater the effect. We have already noted in Chapter 1 what we can now see from Table III, that this charge transfer is accompanied by an increase in the electronegativity of the carbon atom and a decrease in that of the fluorine atom resulting in the possibility of back donation of charge from the p orbitals of the fluorine atom to those of the carbon; this effect as we have seen complicates particularly discussions about the structure of radicals such as R—ĊH—OH.

EXAMPLES

The remainder of this chapter is given over to discussion of a number of interesting radicals in which an attempt is made to predict some of their structures on the basis of the above discussion. Consideration is given to the four factors:

(*i*) Valence state excitation energy;
(*ii*) Unpaired electron energy;
(*iii*) Bond energy;
(*iv*) Electronegativity.

A. Ethynyl HC≡C˙

This radical can be formally written in two σ and one π-configurations,

$$HC\equiv C\ominus \qquad \underset{O}{\overset{H}{\diagdown}}C=C: \qquad H-C=C:$$

(a)   (b)   (c)

Direct experimental evidence as to the structure of ethynyl is inconclusive. E.s.r. spectroscopy[46] gives $a_H = 16$ gauss, a coupling constant which is hard to reconcile with structure (a) but, since its sign is unknown, does not rule our (b) or (c). Turning to our four factors,

(*i*) In structures (b) and (c) one of the carbon atoms is in its ground electronic state while (a), with two carbon atoms in the valence state, is less favoured energetically by an amount which, as we have already noted, could be as much as 100 kcal mol^{-1} or as little as 39 kcal mol^{-1}.

(*ii*) The binding energy of an electron being a function of its s-character, is greatest in (a) where the orbital containing the unpaired electron has 50% s-character and decreases through (b) with 33% to (c) with no s-character. This factor predicts that (a) is more stable than (b) which in its turn is favoured over (c). However it seems possible that overlap between the half-filled p orbital in (c) with an empty p orbital on the adjacent ground state carbon atom could reduce the energy of the free electron by delocalization. Delocalization seems unlikely in (a) and (b).

(*iii*) Structure (a) has the highest bond energy with a C≡C triple bond and a C_{sp}—H σ-bond. As we shall see in Chapter 3, the energy required to break these bonds without causing change in the electronic structure of the atoms joined is about 340 kcal mol^{-1}. The analogous energies of the C—H bonds in (b) and (c) are about 110 and 130 respectively but it is not yet possible to put a figure on the strengths of the C=C bonds of these structures. However, it is not unreasonable to guess that these would be less than about 170 kcal mol^{-1}, the strengths of conventional carbon-carbon double bonds. Taken as a criterion, therefore, bond energy considerations make (a) more stable by at least 40 kcal mol^{-1} over (c) and at least 60 kcal mol^{-1} over (b).

(*iv*) In (a) both carbon atoms will have similar electronegativities, that attached to the hydrogen atom having slightly less due to charge displacement from the hydrogen atom. In (b) and (c) the ground state carbon atom has a low electronegativity (Table III) and since sp bonded carbon is more electronegative than sp^2 bonded carbon there will be more charge separation and therefore greater stability in (c) than in (b).

The conclusions to be drawn from this discussion on the structure of the ethynyl radical hinge on the magnitude of the valence state excitation energy for carbon. If the higher value of about 100 kcal mol^{-1} obtains then this is probably the most significant contribution to the total energy equation and militates against structure (a). If on the other hand its value is 39 kcal mol^{-1} then the analysis is less clear cut. From the combination of factors (i) and (iii) to which we have been able to put some sort of figures, no clear preference emerges between (a) and (c) while (b) is less favoured. Delocalization (ii) and electronegativity (iv) effects help stabilize structure (c) while (a) is favoured by the greater binding energy of the free electron (ii). No distinction can be drawn between these two structures on the basis of these arguments but it is instructive to look at the example of the vinyl radical which we can again depict in extreme forms, the σ (d) and the π (e):

(d)   (e)

In vinyl factor (i) is inapplicable and there is presumably no delocalization. Also, electronegativity effects are likely to be small so that whichever structure is the more stable is determined by, on the one hand, the binding energy of the electron and on the other the energy of the bonds in the radical. The latter factor, as can be seen from Table I in Chapter 3, favours (e) by about 20 kcal mol^{-1} but this is opposed by the greater binding energy of the electron in structure (d) and e.s.r. spectroscopy shows that the latter is the predominant factor and the vinyl radical in the ground state is σ with a structure close to (d). However there are indications that the extent of stabilization of (d) relative to (e) is not large since the introduction of a methyl group to give methylvinyl, $CH_3\dot{C}=CH_2$ appears to stabilize the π structure.[47] As we shall see in Chapter 3 the stabilizing effect of a methyl group is small. Assuming that the difference in free electron binding energy between structures (a) and (c) of the acetylene molecule is not much greater than that between (d) and (e) of the vinyl radical and bearing in mind the stabilization of structure (c) to factors (ii) and (iv) and delocalization, it is suggested that, with a value of 39 kcal mol^{-1} for the valence state excitation energy of carbon, ethynyl is most closely represented by the elementary structure (c). If this is correct and the e.s.r. spectrum of (c) is explainable in terms of the McConnell equation,[6] then with $Q = -24$ gauss, the unpaired spin density on the carbon atom formally associated with the free electron is 0·67. Structure (c) would *a fortiori* be the more stable structure if the valence state excitation energy of carbon is taken as $100 \, \text{kcal mol}^{-1}$. As we shall see in Chapter 5, chemical evidence also favours structure (c).

B. Fluorinated Methyl Radicals, $CF_3^{\cdot}$, $CHF_2^{\cdot}$ and $CH_2F^{\cdot}$ [48]

As we have noted above methyl is more stable in the planar form, factors (*iii*) and (*iv*) predominating over factor (*ii*) but on replacing the hydrogen atoms by the highly electronegative fluorine atoms maximum charge separation (*iv*) is favoured by a reduction in the s-character of the bonds between the trigonal carbon atom and the fluorine atoms. $CH_2F^{\cdot}$ is more or less planar and sp^2 bonded but the deviation from planarity increases through $CH_2^{\cdot}$ (15°) to $CF_3^{\cdot}$ which is sp^3 bonded and tetrahedral.

C. Chloromethyl Radicals

Of this group of radicals the most interesting is $CCl_3^{\cdot}$. The chlorine atom is relatively electronegative and purely on the basis of factor (*iv*) we might expect this radical to be tetrahedral. But b (Table III) is large and the electronegativity of the chlorine atom will decrease with charge displacement from the carbon which in its turn becomes more electronegative. It may be that the most stable configuration of this radical is the one in which the bonding arrangement around the carbon atom is sp^2 and the p orbital containing the unpaired electron overlaps with p orbitals on the chlorine atoms containing unshared pairs of electrons giving rise to back donation, thus,

Spectroscopic measurements[49] on the $CCl_3^{\cdot}$ radical obtained from rapid freezing from the gas phase show the radical to be tetrahedral and this structure is consistent with the rate constant which has been obtained for recombination of two $CCl_3^{\cdot}$ radicals in the gas phase[50] (see p. 97).

Recent work[53] has shown the radical to be bent in solution and to have a structure intermediate between $CH_3^{\cdot}$ and $CF_3^{\cdot}$. But the recombination rate constant in the liquid phase is characteristic of π-radicals.[51] Clearly any attempt to reconcile these observations at this stage is pure speculation but it is possible that the radical is stable in the planar form but that when produced in the gas phase under the conditions used in refs. 49 and 50 it does not readily relax to its ground state, perhaps because of a high barrier to inversion.

D. Hydroxymethyl Radicals

The hydroxyl group is also highly electronegative and the deviation of the α-hydrogen atom coupling constant of the radical $R\dot{C}H(OH)$ (where R is alkyl) from that expected from application of the McConnell equation[18] and the equation for the ^{13}C coupling constant has been interpreted as resulting from a deviation from planarity in this radical. The radical $H\dot{C}(OH)_2$ has $a_{\alpha\text{-H}}$ close to that expected from a planar radical[21] but then so has $\dot{C}HF_2$ which is markedly non-coplanar and it seems plausible that $a_{\alpha\text{-H}}$ is positive and $H\dot{C}(OH)_2$ non-planar.

E. Carbonyl Radicals, $R\dot{C}O$

Formyl can be written in two extreme forms,

$$H-\overset{\displaystyle(\cdot)}{\underset{\displaystyle()}{C}}=O \qquad\qquad \overset{\displaystyle H}{\diagdown}\underset{\displaystyle()}{C}=O$$

(a)   (b)

and the effects of our four factors summarized thus:

 (*i*) Inapplicable.
 (*ii*) The electron is more strongly bound in (b) but in (a) there is the possibility of conjugation with p orbitals on the oxygen atom.
 (*iii*) The bond energy is greater in (a).
 (*iv*) Greater charge separation is possible in (b).

In fact experimental evidence shows formyl to be bent[16] and the same appears to be true of acetyl and benzoyl.[52] It seems that any stabilization due to conjugation or hyperconjugation effects of the benzene ring or methyl is not sufficient to outweigh the combined effects of (*ii*) and (*iv*).

F. Alkoxy Radicals, RO˙

In the parent alcohol, as in water, the electrons of the oxygen atom are arranged so as to include some 2s character in the bonding, i.e. the oxygen atom has been excited to the valence state in order to increase the bond strength and also to reduce steric crowding since the introduction of s-character into the bonds allows the bond angles to open out to more than the 90° of pure p orbitals. Removal of the alcohol hydrogen atom gives a radical either without change in the overall electronic disposition (a) or a π-radical (b) where, to a first approximation, the oxygen atom is in its ground state and the bonding is p.

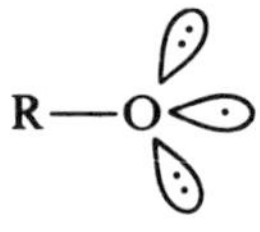

(a)    (b)

Applying the four principles:

(*i*) favours (b);
(*ii*) the electron is more firmly held in (a) but there is the possibility of conjugation or hyperconjugation in (b);
(*iii*) favours (a);
(*iv*) favours (a).

It seems that the largest of these factors is (*i*) and e.s.r. evidence shows that alkoxy and even more, aryloxy, radicals are π. Similar arguments apply to amino radicals.

G. Electronically Excited Radicals

So far when excited radicals have been mentioned, the excess energy has been assumed to be in the vibrational or rotational modes but it is also possible to excite a radical to the σ-state electronically as, for example where one of a lone pair of electrons in an orbital with some s-character is promoted to a half-filled p or π orbital on the atom carrying the unpaired electron. This has been suggested as an explanation of the observed reactivity of DPPH under u.v. light,[5]

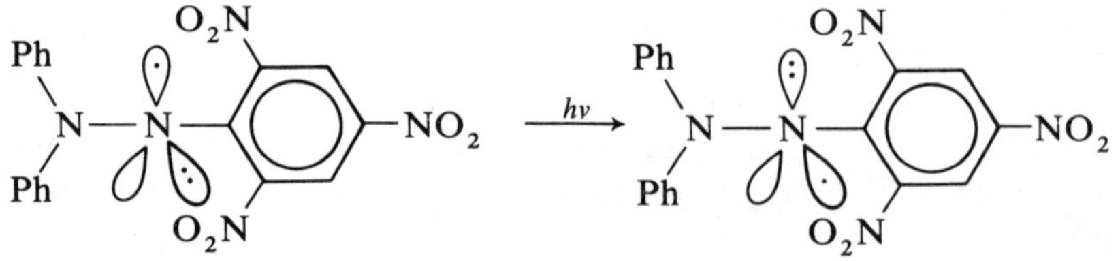

The site of the unpaired electron, which is localized, is more reactive. Some further reactions of excited radicals will be discussed below.

REFERENCES

1. Porter, G. and Ward, B. (1965). *Proc. Roy. Soc. A*, **287**, 457.
2. Porter, G. (1967). *In* "Photochemistry and Reaction Kinetics", Cambridge University Press; Willets, F. W. (1971) *Progress in Reaction Kinetics*. **6**, 52.
3. Bamford, C. H. and Tipper, C. F. H. (eds.) (1969) "Comprehensive Chemical Kinetics", 2, Elsevier, New York.
4. (a) Ingram, D. J. E. (1958). "Free Radicals", Butterworths, London.
   (b) Buchachenko, A. L. (1965). "Stable Radicals", English translation, Consultants Bureau, New York.
   (c) Assenheim, H. M. (1966). "Introduction to Electron Spin Resonance", Hilger, London.
   (d) Bersohn, M. and Baird, J. C. (1966)."Introduction to Electron Paramagnetic Resonance", Benjamin, New York.
   (e) Poole, C. P. (1967). "Experimental Techniques in Electron Spin Resonance", Wiley, New York.
   (f) *Lab. Pract.* November 1964.
   (g) Altshuler, S. A. and Kozyrev, B. M. (1964). "Electron Paramagnetic Resonance", English Translation, Academic Press, New York.
   (h) Fraenkel, G. K. (1960). *In* "Techniques of Organic Chemistry", Vol. 1, pt. 4, Interscience, New York.
   (i) Bersohn, R. (1962). *In* "Determination of Organic Structures by Physical Methods" (F. C. Nachod and W. D. Phillips, eds.), Vol. 2, Academic Press, New York.
   (j) Symons, M. C. R. (1963). *In* "Advances in Physical Organic Chemistry" (V. Gold, ed.), Vol. 1, Academic Press, London.
   (k) Carrington, A. (1963). *Quart. Rev.* **17**, 67.
   (l) Carrington, A. and McLachlan, A. D. (1967). "Introduction to Magnetic Resonance", Harper, New York.
5. Forrester, A. R., Hay, J. M. and Thomson, R. H. (1968). "Organic Chemistry of Stable Free Radicals", Academic Press, London.
6. McConnell, H. M. (1956). *J. Chem. Phys.* **24**, 764; Bersohn, R. J. (1956). *Chem. Phys.* **24**, 1066; Weissman, S. J. (1956). *J. Chem. Phys.* **25**, 890; Jarrett, H. S. (1956). *J. Chem. Phys.* **25**, 1289.
7. Fischer, H. (1965). *Z. Naturforsch.* 1965, **20Å**, 428
8. Levy, D. H. (1965). *Mol. Phys.* **10**, 233.
9. Colpa, J. P. and De Boer, E. (1964). *Mol. Phys.* **7**, 333; Colpa, J. P., De Boer, E., Lazdius, D. and Karplus, M. (1967). *J. Chem. Phys.* **47**, 3089; Purins, D. and Karplus, M. (1968). *J. Am. Chem. Soc.* **90**, 6275.
10. Hausser, K. H., Brunner, H. and Jochins, J. C. (1966). *Mol. Phys.* **10**, 253.
11. Sogo, P. B., Nakazaki, M. and Calvin, M. (1957). *J. Chem. Phys.* **26**, 1343.
12. Carrington, A. (1963). *In Quart. Rev.* **17**, 67.
13. Karplus, M. and Fraenkel, G. K. (1961). *J. Chem. Phys.* **35**, 1312.
14. Melchior, M. T. and Maki, A. H. (1961). *J. Chem. Phys.* **34**, 471; Carrington, A. and Santos-Viega, J. (1962). *Mol. Phys.* **5**, 21; Rieger, P. H. and Fraenkel, G. K.

(1962). *J. Chem. Phys.* **37**, 2795; (1963). **39**, 609; Strom, E. T. Russell, G. A. and Konaka, R. (1965). *J. Chem. Phys.* **42**, 2033; Henning, J. C. M. (1966). *J. Chem. Phys.* **44**, 2139.

15. Kaplan, M., Bolton, J. R. and Fraenkel, G. K. (1965). *J. Chem. Phys.* **42**, 955.
16. Adrian, F. J. Cochran, E. L. and Bowers, V. A. (1962). *J. Chem. Phys.* **36**, 1661.
17. Lau, P. W. and Lin, W. C. (1969). *J. Chem. Phys.* **51**, 5139.
18. Fessenden, R. W. (1967). *J. Phys. Chem.* **71**, 74; Shiga T., Boukhors A. and Douzou, P. (1967). *J. Phys. Chem.* **71**, 3559, 4264.
19. Dixon, W. T. and Norman, R. O. C. (1963). *J. Chem. Soc.* 3119.
20. Fischer, H. (1964). *Z. Naturforsch.* **19A**, 866; (1965) **20A**, 428.
21. Dobbs, A. J., Gilbert, B. C. and Norman, R. O. C. (1971). *J. Chem. Soc. A*, 124.
22. Webb, G. A. (1970). *In Ann. Repts. NMR Spectroscopy* **3**, 211.
23. De Boer, E. and MacLean, C. (1966). *J. Chem. Phys.* **44**, 1334.
24. Bloembergen, N. (1957). *J. Chem. Phys.* **27**, 595; McConnell, H. M. and Chesnut, D. B. (1958). *J. Chem. Phys.* **28**, 107.
25. Hausser, K. H., Brunner, H. and Jochins, J. C. (1965). *Mol. Phys.* **10**, 253; Kreilick, R. W. (1966). *J. Chem. Phys.* **45**, 1922.
26. Andrews, L. and Pimentel, G. C. (1967). *J. Chem. Phys.* **47**, 3637.
27. Shirk, J. S. and Pimentel, G. C. (1968). *J. Am. Chem. Soc.* **90**, 3349.
28. Bennett, J. E., Mile, B. and Thomas, A. (1969). *Proc. Roy. Soc. A*, **293**, 246.
29. Fessenden, R. W. and Schuler, R. H. (1965). *J. Chem. Phys.* **43**, 2704.
30. Lankamp, H., Nauta, W. Th. and MacLean, C. (1968). *Tet. Lett.* 249.
31. Coulson, C. A. (1961). 2nd edition. "Valence", Oxford University Press.
32. Hay, J. M. (1970). *J. Chem. Soc. B*, 45.
33. Fessenden, R. W. and Schuler, R. H. (1963). *J. Chem. Phys.* **39**, 2147.
34. Herzberg, G. (1961). *Proc. Roy. Soc. A*, **262**, 291.
35. Chang, S. Y., Davidson, E. R. and Vincow, G. (1970). *J. Chem. Phys.* **52**, 5596.
36. Walsh, A. D. (1953). *J. Chem. Soc.* 2288, 2306, 2325.
37. Kibby, C. L. and Weston, R. E. Jr. (1968). *J. Am. Chem. Soc.* **90**, 1084.
38. Drago, R. S. and Peterson, H. (1967). *J. Am. Chem. Soc.* **89**, 5774.
39. Dixon, W. T. and Norman, R. O. C. (1964). *J. Chem. Soc.* 4857.
40. Streitwieser, A. Jr. (1961). "Molecular Orbital Theory for Organic Chemists", Wiley, New York.
41. Freed, J. H. (1965). *J. Chem. Phys.* **43**, 1710.
42. Hedaya, E., Hinman, R. L., Schomaker, V., Theodoropulos, S. and Kyle, L. M. (1967). *J. Am. Chem. Soc.* **89**, 4875.
43. Cyr, N. and Lin, W. E. (1967). *Chem. Comm.* 192.
44. Walter, R. J. (1966). *J. Am. Chem. Soc.* **88**, 1923.
45. Huheey, J. E. (1964). *J. Phys. Chem.* **68**, 3073; (1965) **69**, 3284; (1966). **70**, 2086.
46. Cochran, E. L., Adrian, F. J. and Bowers, V. A. (1964). *J. Chem. Phys.* **40**, 213.
47. Hay, J. M. and Lyon, D. (1970). *Proc. Roy. Soc. A*, **317**, 21.
48. Fessenden, R. W. and Schuler, R. H. (1965). *J. Chem. Phys.* **43**, 2704.
49. Andrews, L. (1968). *J. Chem. Phys.* **48**, 972.
50. Andrews, L. (1967). *J. Phys. Chem.* **71**, 2761.
51. Tedder, J. M. and Watson, R. A. (1966). *Trans. Faraday Soc.* **62**, 1215.
52. Solly, R. K. and Benson, S. W. (1971). *J. Am. Chem. Soc.* **93**, 1592.
53. Cooper, J., Hudson, A. and Jackson, R. A. (1972). *Mol. Phys.* **23**, 209.
54. Hesse, C. and Roncin, J. (1970). *Mol. Phys.* **19**, 803.

CHAPTER 3
BOND DISSOCIATION ENERGIES

With the exception of the, fairly unusual, reactions of electron or charge transfer, free radical reactions involve the making and/or breaking of chemical bonds. The breaking of a chemical bond can occur in, broadly speaking, two ways, the homolytic,

$$R—R' \rightarrow R^{\cdot} + R'^{\cdot}$$

to give free radicals, and the heterolytic,

$$R—R' \rightarrow R^{+} + R'^{-}$$

producing ions or perhaps radical ions. The making of the bond is, of course the reverse of these processes. In this book we shall confine ourselves to the first of these reactions and in order to complete our basic picture of the factors influencing radical reactivity it is necessary to turn now to a discussion of the strengths of chemical bonds.

The dissociation energy of a chemical bond is a measure of the force between the two atoms constituting the bond. In the simplest case the variation of the potential energy (E_{AB}) of the diatomic molecule, AB, with changing interatomic distance r_{A-B} can be illustrated as in Fig. 1.

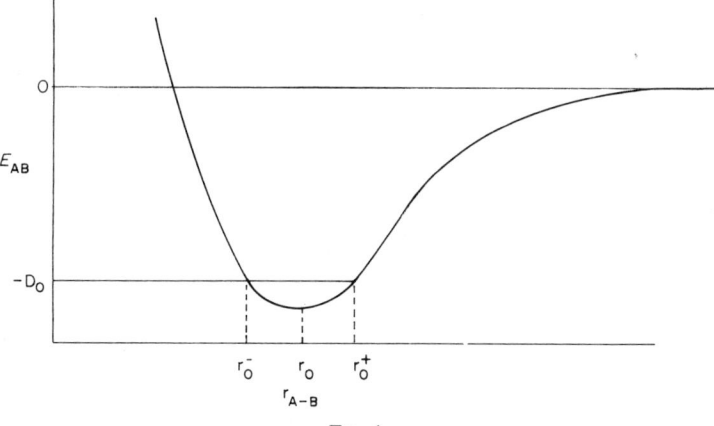

FIG. 1.

The potential energy reaches a minimum at an interatomic separation r_0 but in fact since the molecule in its ground state contains a zero point half

59

quantum of energy $= \frac{1}{2}hv$ (where v is the vibrational frequency of the bond A—B and h is Planck's constant), the ground state molecule vibrates between $r_{A-B} = r_0^-$ and r_0^+ and the interatomic separation, r_0, can be regarded as the average. Reduction of the interatomic distance significantly below r_0 produces a rapid increase in potential energy simply discussed in terms of the close approach of two shells of negatively charged electrons. Increase in r_{A-B} ultimately results in dissociation of the molecule when the atoms are infinitely far apart or, more practically, at such a distance apart that the forces between them are negligible. The energy difference between this state and that of lowest potential energy is the dissociation energy of the molecule at absolute zero of temperature with all the species considered to be ideal gases, D_0. The bond dissociation energy (b.d.e.), $D(A$—$B)$ is an experimental quantity and is referred for convenience to the standard state at 25°C. $D(A$—$B)$ then differs from $D_0(A$—$B)$ by the enthalpy change associated with a rise in temperature from 0° to 298°K, given at constant pressure by,

$$\Delta H = C_p \Delta T \tag{1}$$

For the dissociation of polyatomic molecules, for example that of ethane into two methyl radicals, the situation is much more complex but it has been summarized conveniently by Benson[1] with the aid of the diagram in Fig. 2.

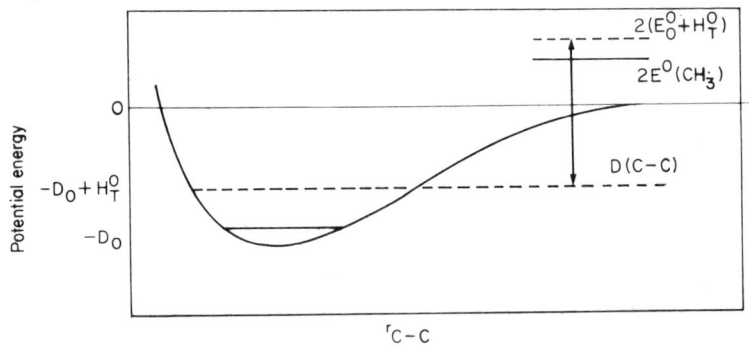

FIG. 2.

H_T^0 is the enthalpy associated with the temperature rise from 0° to 298°K and takes account both of the ethane molecule and of the methyl radicals. E_0° is the zero point energy of the methyl radical. The bond dissociation energy of ethane, giving two methyl radicals is given by,

$$D(C—C) = +D_0 + H_T + 2E_0^\circ(CH_3^\bullet)$$
$$= 49\cdot2 + 2\cdot4 + 2 \times 18\cdot2$$
$$= 87\cdot8 \text{ kcal mol}^{-1}$$

As stated above these definitions of the bond dissociation energy require the reactants and products to be in their ground states so that in the dissociation of the central bond of ethane the process is accompanied by the formation of planar methyl radicals; in other words the hybridization of the carbon atoms changes from sp^3 to sp^2.

EXPERIMENTAL METHODS OF DETERMINING B.D.E'S

The methods used to obtain b.d.e's of molecules have been discussed in considerable detail elsewhere and for the purposes of this book a short outline only of the most important methods will be given.

(a) *Thermodynamic Methods*
For the equilibrium,

$$A - B \rightleftharpoons A + B$$

the equilibrium constant, is

$$K = \frac{[A][B]}{[A - B]} \tag{2}$$

and the free energy change is given by,

$$\Delta G^\circ = -RT \ln K = \Delta H^\circ - T\Delta S^\circ \tag{3}$$

If the equilibrium constant is measured at two temperatures,

$$\Delta H^\circ = \frac{RT_1T_2}{(T_1 - T_2)} \ln (K_1/K_2) \tag{4}$$

From either a knowledge of ΔS° or from a measurement of K at two temperatures and the van't Hoff equation we can obtain ΔH°. The method is limited by the accuracy and sensitivity of the methods used to obtain the equilibrium constant but spectroscopic techniques—both optical and magnetic—are often suitable.

(b) *Spectroscopic Methods*
It is possible to obtain a good estimation of the bond dissociation energy of a diatomic molecule from an analysis of its electronic absorption or emission spectrum. From observations of the electronic energy levels, the potential

energy curve of the molecule (Fig. 1) can be extrapolated to the point at which the potential energy associated with binding is to all intents and purposes zero. In the calculations it is of course necessary to take into account any excess energy, particularly electronic, in the atoms produced in the dissociation. The method has been used for H_2, O_2, F_2, Cl_2, N_2, S_2, NO and CO.

The spectra of molecules containing more than two atoms are so complex as to make this approach impracticable but in favourable cases the use of spectroscopic methods may allow limits to be set on bond energies in polyatomic molecules. For example Hartley[2] found that hydrogen when dissociated in a discharge tube and allowed to react with formaldehyde produces a greenish luminescence characteristic of glyoxal which is presumed to result from the sequence,

$$H^{\cdot} + HCHO \rightarrow H_2 + {}^{\cdot}CHO$$

$$2\,{}^{\cdot}CHO \rightarrow (CHO)_2^{*}$$

The luminescence has a long wavelength limit of 3965 Å which gives an estimate for the upper limit of the bond energy of the glyoxal formed, i.e. $D(OHC—CHO) = 71.5 \pm 0.3\ \text{kcal mol}^{-1}$. From this estimate it is possible to work out bond energy data for formyl and formaldehyde, as follows,

$$\Delta H_f(\dot{C}HO) = \tfrac{1}{2}(D(OHC—CHO) - \Delta H_f(CHO)_2)$$

$$= \tfrac{1}{2}(71.5 - 72.7)$$

$$= 0.6\ \text{kcal mol}^{-1}$$

For the reaction,

$$^{\cdot}CHO \rightarrow CO + H^{\cdot}$$

we can calculate the bond dissociation energy of formyl, thus,

$$\Delta H = D(H—CO) = \Delta H_f(CO) + \Delta H_f(H^{\cdot}) - \Delta H_f({}^{\cdot}CHO)$$

$$= -26.5 + 52.1 + 0.6$$

$$= 26.3\ \text{kcal mol}^{-1}$$

and similarly for,

$$HCHO \rightarrow H^{\cdot} + {}^{\cdot}CHO$$

$$\Delta H = D(H—CHO) = \Delta H_f({}^{\cdot}CHO) + \Delta H_f(H^{\cdot}) - \Delta H_f(HCHO)$$

$$= -0.6 + 52.1 + 27.7$$

$$= 79.2\ \text{kcal mol}^{-1}$$

(c) Photochemical Methods

Closely allied to the spectroscopic methods are those in which photochemical techniques are used to obtain bond energy data. These are based essentially on attempts to find the minimum light energy which will initiate a particular reaction. The wavelength incident on a particular reaction system is increased or decreased by the use of suitable filters until that at which the reaction is either just detectable or just not detectable.

Again we will use as an example the case of formaldehyde, the photochemical decomposition of which can occur formally in one of two ways,

$$HCHO \xrightarrow{h\nu} H^{\cdot} + {}^{\cdot}CHO \tag{1}$$

$$\xrightarrow{h\nu} H_2 + CO \tag{2}$$

Since the former reaction is expected to be followed by the abstraction reaction,

$$H^{\cdot} + HCHO \rightarrow H_2 + {}^{\cdot}CHO \tag{3}$$

both mechanisms, produce molecular hydrogen, the detection of which is used as an indication of reaction taking place. But a distinction between the two processes can be made by photolyzing mixtures of formaldehyde and deuterated formaldehyde; if molecular hydrogen is a product of the radical process, then HD should be detectable, i.e.

$$H^{\cdot} + DCDO \rightarrow HD + {}^{\cdot}CDO$$

$$D^{\cdot} + HCHO \rightarrow HD + {}^{\cdot}CHO$$

The molecular process, reaction (2) on the other hand should produce no isotopically mixed hydrogen. In the event mixed hydrogens were detected[?] and it was found that the free radical dissociation of a hydrogen atom, reaction (1) requires an incident wavelength greater than 3650 Å. This sets a lower limit on $D(H—CHO)$ of 78·3 kcal mol^{-1}, a value close to that derived above. It should be noted that formaldehyde provides a particularly sensitive test of this approach since the production of, in theory, just one hydrogen atom leads to a chain reaction and the production of molecular hydrogen in detectable quantities, thus,

$$H^{\cdot} + HCHO \rightarrow H_2 + {}^{\cdot}CHO$$

$${}^{\cdot}CHO \rightarrow CO + H^{\cdot}$$

A still more complex example is that of methanol for which one can write down a number of possible photochemical decompositions:

$$CH_3OH \xrightarrow{h\nu} CH_3O^{\cdot} + H^{\cdot} \tag{4}$$

$$\rightarrow \cdot CH_2OH + H^\cdot \tag{5}$$

$$\rightarrow H_2 + HCHO \tag{6}$$

$$\rightarrow CH_2 + H_2O \tag{7}$$

Step (7) is ruled out at wavelengths greater than 1800 Å. At 1800 Å experiments with mixtures of CD_3OH and CH_3OD followed by analysis of the products showed[4] that the major primary process is (7) and that almost all the residual reaction was (4), demonstrating that,

$$D(CH_3O - H) < D(H - CH_2OH)$$

At even shorter wavelengths emission can be observed[5] from hydroxyl produced in the dissociation,

$$CH_3OH \xrightarrow{h\nu} CH_3^\cdot + \cdot OH(^2\Sigma^+)$$

In general it will be obvious that photochemical methods of determining bond dissociation energies are also subject to the requirement that the electronic energy of the fragments must be known and any excess taken into account in the calculations.

(d) Electron Impact Methods[6]

If, as the result of electron impact, an ion is produced in the reaction,

$$XY + e^- \rightarrow X^+ + Y + 2e^-$$

we can write the appearance potential of the X^+ ion as,

$$V(X^+) = D(XY) + I(X) + KE + EE$$

where $I(X)$ is the ionization potential of the atom X and the last two terms give respectively the excess kinetic and excess electronic energies of the ion and/or the neutral atom Y. A knowledge of the values of the last three terms on the right-hand side of the equation leads to a value for $D(X—Y)$. In fact the kinetic energies can be measured qualitatively or even semi-quantitatively but the simplest situation arises where the ion and neutral fragment are produced without excess kinetic energy and in their ground states, for example in the process,

$$CH_4 + e^- \rightarrow CH_3^+ + H^\cdot + 2e^- \quad V = 14\cdot39 \text{ eV}$$

A knowledge of $I(CH_3^\cdot)$ measured in a separate experiment (9·56 eV) gives,

$$D(H_3C—H) = 14\cdot39 - 9\cdot86 = 4\cdot53 \text{ eV}$$

$$= 104\cdot4 \text{ kcal mol}^{-1} \quad (1 \text{ eV} = 23\cdot05 \text{ kcal})$$

Where the ionization potential of the radical is not known an indirect approach may be tried. For example suppose we require $D(H_3C—C_2H_3)$. We can perform the processes,

$$H_2 + e^- \rightarrow H^+ + H^{\cdot} + 2e^- + V_1$$

$$C_2H_3—H + e^- \rightarrow H^+ + C_2H_3^{\cdot} + 2e^- + V_2$$

in the mass spectrometer, obtaining the appearance potential of H^+ in each case. Making the assumption that V_1 and V_2 contain no kinetic energy terms, we get by subtraction,

$$H_2 - C_2H_4 = H^{\cdot} - C_2H_3^{\cdot} + V_1 - V_2 \qquad (8)$$

If in addition we know the heats of formation of H_2, $CH_3—C_2H_3$, C_2H_4 and CH_4 and also the dissociation energy $D(CH_3—H)$, then we can calculate ΔH for the reaction,

$$H_2 + CH_3—C_2H_3 \rightarrow C_2H_4 + H^{\cdot} + CH_3^{\cdot} + \Delta H \qquad (9)$$

and now, by subtracting equation (8) from (9) we get,

$$CH_3—C_2H_3 \rightarrow CH_3^{\cdot} + C_2H_3^{\cdot} + \Delta H + V_1 - V_2$$

whence

$$D(CH_3—C_2H_3) = \Delta H + V_1 - V_2$$

As in this case it is usually arranged that one bond energy, i.e. $D(CH_3—H)$ is known and from this a series of others may be derived. As the number of known free radical ionization potentials has increased this indirect method has become less frequently used but it still has applications as a confirmatory technique.

Partly because of the difficulties arising from the unknown states of the fragments produced in dissociation under electron impact and partly because of the possibility of complex dissociation processes from excited states of molecules and fragments, the results of bond dissociation energy determinations by electron impact methods have often disagreed markedly from those obtained by other methods but since excess energy is usually ignored the results have generally erred on the high side. If the fragment ions come off with less than equilibrium translational or rotational energies then the measured bond dissociation energy might be low but not by more than about 2 kcal mol^{-1}.

(e) Kinetic Methods

Many of the bond dissociation energies commonly used have been obtained from detailed studies of chemical reactions. Many approaches have been devised but we will mention two only here.

(i) *The unimolecular decomposition of a molecule.* If the activation energy of the reaction,

$$R_1-R_2 \rightarrow R_1^{\cdot} + R_2^{\cdot} \qquad (10)$$

can be measured from a study of the unimolecular decomposition of the molecule R_1R_2 over a range of temperature, then making the usual assumption (see Chapter 4) that the activation energy of the reverse reaction, the recombination of the radicals, is zero, E_{10} is the heat of the reaction and the required bond dissociation energy. It is, of course, necessary that the measured rate of disappearance of the reactant is due entirely to the unimolecular dissociation and that its disappearance by other pathways such as radical attack is negligible. This criterion must be established by the sort of detailed kinetic and analytical studies such as those carried out on di-t-butyl peroxide,[7] where the mechanism has been established to be, $(t\text{-Bu} = (CH_3)_3C-)$

$$t\text{-BuO}_2 t\text{-Bu} \rightarrow 2t\text{-BuO}^{\cdot}$$

$$t\text{-BuO}^{\cdot} \rightarrow CH_3COCH_3 + CH_3^{\cdot}$$

$$2CH_3^{\cdot} \rightarrow C_2H_6$$

$$CH_3^{\cdot} + CH_3COCH_3 \rightarrow CH_4 + CH_3CO\dot{C}H_2$$

$$CH_3^{\cdot} + {}^{\cdot}CH_2COCH_3 \rightarrow C_2H_5COCH_3$$

Broadly the reaction proceeds about 90% by the overall process,

$$t\text{-BuO}_2 t\text{-Bu} \rightarrow 2CH_3COCH_3 + C_2H_6$$

and about 10% by,

$$t\text{-BuO}_2 t\text{-Bu} \rightarrow CH_3COCH_3 + C_2H_5COCH_3 + CH_4$$

Another approach is to use the properties of the shock tube to study the decomposition at very short reaction times before secondary reactions can occur to complicate the process and the calculation of the results. Alternatively the principle of radical trapping can be used to remove, by the use of scavengers such as benzyl, radicals R_1 and/or R_2 which are sufficiently reactive to attack the parent molecule. This is the basis of the well-known toluene carrier technique[8] in which the main sequence of reactions is,

$$R_1R_2 \rightarrow R_1^{\cdot} + R_2^{\cdot}$$

$$R_1^{\cdot}(R_2^{\cdot}) + C_6H_5CH_3 \rightarrow R_1H(R_2H) + C_6H_5CH_2^{\cdot}$$

$$2C_6H_5CH_2^{\cdot} \rightarrow (C_6H_5CH_2)_2$$

Aniline and hydrogen iodide have also been used as scavengers.

(ii). A second kinetic approach is to carry out a detailed study on (preferably) a simple radical reaction, for example the reaction between iodine and an organic iodide which has been shown by Benson and Golden[9] to go quantitatively at measurable rates in the temperature range 200°–500°C via the following probable mechanism,

$$I_2 + M \underset{v}{\overset{p}{\rightleftharpoons}} 2I^\bullet + M \quad k_{I_2}$$

$$RI + I^\bullet \underset{2}{\overset{1}{\rightleftharpoons}} R^\bullet + I_2$$

$$R^\bullet + HI \underset{4}{\overset{3}{\rightleftharpoons}} RH + I^\bullet$$

If the entropy and heat of formation of (a) RI or (b) RH are known, values for the heat of formation and entropy of $R^\bullet$ can be derived from measurements of (a) $\Delta H^\circ_{1,2}$ and $\Delta S^\circ_{1,2}$ or (b) $\Delta H_{3,4}$ and $\Delta S_{3,4}$. Thus if the energies of activation of one of the pairs of reactions 1,2 or 3,4 is measurable so also is the heat of formation of the radical. Since E_2 and E_3 are fairly well established at 0 ± 1 and $1 \pm 1 \text{kcal mol}^{-1}$ respectively, measurements of E_1 and E_4 in kinetic experiments will give the heat of formation of the radical and hence the bond dissociation energy required.

(f) Reactions of Hot Recoil Tritium Atoms

Tritium atoms produced in nuclear recoil reactions have been shown to abstract hydrogen atoms from organic substrates,[10]

$$T^* + RH \rightarrow HT + R^\bullet$$

At high energies the abstraction process appears to be one of stripping in which only a small fraction of the energy of the hot tritium atom is available to further the abstraction process,

The reaction rate therefore becomes rather critically dependent on the strength of the bond being broken and even small differences in the relevant bond dissociation energies become important; weaker bonds, in promoting abstraction by the high energy stripping process, raise the average energy for abstraction.

When the tritium atoms are moderated by performing the abstraction reaction in the presence of excess of perdeuterioethylene or fluorinated hydrocarbons the stripping process appears to be suppressed and a correlation can be found between the yield of HT and the bond dissociation energy of the substrate, i.e. $D(R—H)$. It is not yet possible to describe the abstraction reaction in detail since the loss in energy of the tritium atoms, which is a necessary precondition for the successful formation of bonds with these atoms, is a somewhat random process and must be critical at and just above the energy of bond formation. But relative yields of HT from different hydrocarbons are comparable under equivalent reaction conditions and, when perfluorocyclobutane is used as moderator a smooth curve relates these to the bond dissociation energies of the bonds broken to within quite close limits.

Later work[11] has, however, shown that the smoothness of the correlation obtained in the case of a series of hydrocarbons may be misleading when the approach is applied to different series of molecules, for example halogenated methanes. It was found that, using the correlation previously established for hydrocarbons, the measured yields of HT predicted values for $D(CF_3—H)$ and $D(CF_2H—H)$ which are much lower than those obtained by other methods. The authors' explanation, that the discrepancy is connected with residual energy in the organic fragment, is identical to that used in Chapter 2 to discuss molecular dissociations. Since the equilibrium configuration of $CF_3^•$ differs little if at all from that in the CF_3H molecule, there is no significant change in geometry on the formation of the radical by hydrogen abstraction. In other words there is no reason to suppose that the product $CF_3^•$ radical is vibrationally excited by the process. The picture is quite different, however, in the formation of hydrocarbon radicals in the same process since these are planar about the trigonal carbon atom in contrast to their tetrahedral shape in the parent molecule. Hydrogen abstraction, then, is accompanied by a change in geometry and it is likely that under the high energy conditions obtaining during this abstraction reaction, the radicals are produced with an excess of vibrational energy, in other words they are produced as σ-radicals. It would seem logical to correlate the results from recoil tritium atom abstraction in the hydrocarbon series with what we shall define below as instaneous bond dissociation energies and it is gratifying to see that extrapolation of the curve in figure 1 of reference 11 gives a value for the instantaneous bond dissociation energy in methane of about 109 kcal mol^{-1}, precisely the value which is derived below.

There are further complications with this particular method of measuring bond dissociation energies. For example relative yields of HT in the presence of different moderators differ not just overall but also in detail. Further work is clearly necessary.

(g) Empirical Methods

In addition to these experimental methods for determining bond dissociation energies, a number of attempts have been made to link other experimental results and b.d.e's usually in a largely empirical manner. In the most simple of these it is assumed that the activation energy of the reaction,

$$X^{\cdot} + RH \rightarrow XH + R^{\cdot} \tag{10}$$

is a function only of the strengths of the bonds being broken and formed, in other words all other effects: for example those associated with polarity, are negligible. In such a case it is not unreasonable to expect that, for constant X, the activation energies of reactions (10) over a series of RH should be a function of the strength of the R—H bond, and further, that if in a series of different radicals X there are similar factors controlling the activation energies of the forward and backward reactions, a wide correlation should exist between the activation energies of the abstraction reactions and the dissociation energy of the bond which is being broken.

Evans and Polanyi[12] were the first to quantify this argument using for exothermic reactions the relationship,

$$\Delta E = -\alpha H \tag{5}$$

With the acquisition of more experimental data, particularly on the reactions of sodium atoms, the equation was later refined to,

$$E_1 = A - \alpha q \tag{6}$$

where A is constant for a given homologous series and, for sodium atoms reacting with alkyl chlorides α = about 0·27.

Semenov[13] further extended this work to include the reactions of a number of radicals such as methyl, sodium, hydrogen and hydroxyl with a range of different substrates. Within the error limits $E = \pm 1$–2 kcal mol^{-1} and $\Delta H = \pm 2$–3 kcal mol^{-1} most of this data can be fitted on to a curve described by the equation,

$$E = 11 \cdot 5 - 0 \cdot 25(\Delta H) \tag{7}$$

For exothermic reactions,

$$E = 11 \cdot 5 - 0 \cdot 25 \, \Delta H \tag{8}$$

and for endothermic reactions,

$$E = 11 \cdot 5 + 0 \cdot 75 \, \Delta H \tag{9}$$

Too much, however, cannot be expected of a general equation and as detailed experimental results have accumulated for a number of different radicals, $X^{\cdot}$, it has been possible to derive relationships for individual radicals or sets of radicals.[14] For chlorine, bromine and iodine atoms for example,

$$E = \alpha \Delta H + c \tag{10}$$

where $\alpha = 0.91$ and $c = 4.3 \, \text{kcal mol}^{-1}$.

Another relationship takes the form,

$$E = \alpha[D(\text{R}—\text{H}) - C] \tag{11}$$

with the following factors for the radicals indicated:

X·	α	C	$D(\text{H}—\text{X})$
I·	0.97	69.0	71.3
NF$_2$·	1.1	76.1	85
Br·	0.86	82.5	87.5
CH$_3$·	0.49	74.3	104

There are a number of objections to this approach. Even within a closely related series of reactions the activation energies of the forward and back reactions (10) are not necessarily controlled by the same mechanism for different X·. Further, the simple formulae predict negative activation energies when applied to sufficiently exothermic reactions. Johnston[15] has shown that some improvement can be introduced by the addition of certain kinetically derived empiricisms but that even better correlations can be obtained by speculating outside the field of chemical kinetics altogether. This author has produced an improved scheme in which the potential energy of the system,

$$\text{X} \underset{r_1}{\quad\rule{1cm}{0.4pt}\quad} \text{H} \underset{r_2}{\quad\rule{1cm}{0.4pt}\quad} \text{R}$$

representing different distances of approach of the three interesting reaction centres in eqn. (10), is given by,

$$V = E(\text{XH}) + E(\text{H}—\text{R}) + E(\text{XR}) \tag{12}$$

and is computed for varying values of r_1 and r_2 with the condition that the total bond order X—H and H—R is unity. It is found that the bonding contribution to V has about the same importance as the repulsive term in determining the value of the activation energy. The latter term in all cases reported by Johnston lies between 2 and 8 kcal mol^{-1} and attains a maximum at equal (half) bond orders. The numerical values of the summation in eqn. (12) go through a maximum, the potential energy of activation energy which, while differing from the experimental activation energies because of differences in the zero point and internal excitation energies of the reactants and the complex, are more realistic for highly exothermic and endothermic processes than those derived from the Polanyi–Semenov equation in that for exothermic reactions the activation energy approaches zero and for endothermic reactions it approaches the heat of the reaction. Unfortunately this approach, while, in most cases, an improvement over the earlier empirical treatments has lost its essential and intuitive simplicity; after all much of the value of simple relationships lies in their predictiveness and not in their predictability. We shall return in Chapter 5 to the question of empirical relationships between activation energies and bond dissociation energies.

MECHANISM OF BOND DISSOCIATION

In Chapter 2 it was suggested that in the case of a dissociation of an organic molecule RR',

$$RR' \rightarrow R^{\cdot} + R^{\cdot \cdot} \tag{11}$$

the application of sufficient energy will produce radicals without change in the hybridization around the carbon atoms linked by the bond being broken. With a smaller amount of energy the two parts of the molecule will rehybridize into their most stable form, in the case of the dissociation of a saturated hydrocarbon involving a change in hybridization around the ultimately trigonal carbon atom from sp^3 to sp^2. The energy released during this rehybridization helps reduce the energy required to rupture the bond to its lowest possible value, the bond dissociation energy (b.d.e.).

It might be conjectured, therefore, from this picture that b.d.e.s obtained by experimental methods might be expected to be affected by the amount of energy available and this has frequently been shown to be the case. We have already mentioned the reactions of hot recoil tritium atoms.

Photolytic and electron impact methods should produce the same sort of effect but the two bond dissociations,

$$HCHO \rightarrow H^{\cdot} + {}^{\cdot}CHO$$
$$C_2H_4 \rightarrow H^{\cdot} + {}^{\cdot}C_2H_3$$

provide apparent exceptions to this generalization in that, although the product radicals are stable in the σ-configuration, the b.d.e's are lower than those derived from kinetic studies.[1, 2, 16] The explanation may lie in a two-step dissociation process; to take formaldehyde, photon impact raises the molecule to an excited vibrational state in which the coplanarity of the molecule is lost, the s-character of one of the C—H bonds reduced and absorption of a quantum of a lower energy is sufficient to break this weakened bond, thus,

$$
\begin{array}{c}
\text{H} \\
| \\
\text{H}-\text{C}=\text{O}
\end{array}
\qquad
\xrightarrow[hv]{76\,\text{kcal}}
\quad
\overset{\bullet}{\text{H}} + \overset{\bullet}{\text{C}}\text{HO}
$$

$$
\uparrow hv
\qquad\qquad
\underset{88\,\text{kcal}}{\overset{hv}{\nearrow}}
$$

$$
\begin{array}{c}
\text{H} \\
\diagdown \\
\qquad\text{C}=\text{O} \\
\diagup \\
\text{H}
\end{array}
$$

While such a sequence may be very rare, it will be recalled that the production of only one hydrogen atom could result in the initiation of a chain reaction and the production of a measurable amount of hydrogen which is the criterion of the method. Similar arguments apply to the electron impact dissociation of ethylene since this method too depends on the detection of the smallest possible amount of hydrogen.

A similar sort of mechanism may be responsible for the observation[17] that while irradiation of solid acetylene by an r.f. discharge in hydrogen produces no radicals detectable by e.s.r., they are formed on simultaneous photolysis and condensation of acetylene, conditions under which vibrational excitation is possible.

In very broad terms it may be said that "high" b.d.e's are observed when dissociation takes place giving normally σ-radicals with a σ structure, while "low" or equilibrium b.d.e's are favoured by stabilization of the resultant radicals in the π-configuration, for example by conjugation or hyperconjugation the latter effect being shown nicely by the reduction in $D(\text{RO}-\text{H})$ on going from water through CH_3OH to C_6H_5OH,

R—H	$D(\text{R}-\text{H})$ kcal mol^{-1}
HO—H	119
CH_3CO_2—H	112
CH_3O—H	102
C_6H_5O—H	85
HO_2—H	90

The reason for the reduction on going from water to hydrogen peroxide is not so obvious but it is not impossible that the reduction could also be the result of stabilization of the hydroperoxy radical by overlap of the orbital containing the unpaired electron with a non-orthogonal filled p orbital on the oxygen atom, thus,

It is interesting to note that the reduction in going from water to hydrogen peroxide, 29 kcal mol^{-1}, is not very different from the difference in $D(R—H)$ between ethylene and formaldehyde, 23 kcal mol^{-1}, where the same effect can also be visualized,[18]

INSTANTANEOUS BOND DISSOCIATION ENERGIES (I.B.D.E'S)[19]

Instantaneous bond dissociation energies are defined as the energy required to dissociate a molecule RR' into two radicals R˙ and R″ which have the same geometric configuration as they had in the original molecule, the only difference being that there is no longer any overlap between the orbitals now containing a free electron; for example the i.b.d.e. of ethane is the energy required to produce two σ-methyl radicals,

We shall now see that it is possible, using values measured for b.d.e's as well as some suggested i.b.d.e's which have been determined, to calculate i.b.d.e's for C—C and C—H bonds.

In order to build up such a body of data it is necessary to start from authenticated i.b.d.e's, for example from experiments where bond energy values have been derived from dissociations at high energies and shown to differ from results on the same bond obtained under more moderate conditions. One probable example of this is the dissociation of cyanogen into two ˙CN radicals, a process which requires about 133 kcal mol^{-1},[20] some 10 kcal mol^{-1} greater than can be calculated from the well established value of $D(H—CN)$.[21] By comparison[1] $ID(HC_2—C_2H)$ has been estimated as 150 kcal mol^{-1} and from this and the heat of formation of diacetylene, 113·0 kcal mol^{-1},[22] we can estimate that the instantaneous bond dissociation energy for the production of σ-ethynyl from acetylene, $ID(C_{sp}—H)$ is, ignoring delocalization energy in the calculation, about 129 kcal mol^{-1}.

Phenyl is a typical σ-radical and the value of $ID(C_{sp^2}-H)$ can be obtained directly from the heat of formation of phenyl iodide ($38\cdot85$ kcal mol^{-1})[22] and the thermodynamics of the reaction between hydrogen iodide and phenyl iodide[23] as $110\cdot7 \pm 2$ kcal mol^{-1}. From this value and the heat of formation of biphenyl ($40\cdot5$ kcal mol^{-1})[24] we obtain, for $ID(C_{sp^2}-C_{sp^2})$, $116\cdot3 \pm 3$ kcal mol^{-1}. The delocalization energy of biphenyl has been calculated[25] to be negligible and can safely be ignored in this calculation.

Turning to vinyl, we have already calculated the heat of the reaction,

$$C_2H_4 \rightarrow \sigma\text{-}C_2H_3{}^{\cdot} + H^{\cdot}$$

to be $110\cdot7$ kcal mol^{-1} and from this and the heat of formation of ethylene ($12\cdot5$ kcal mol^{-1}) we find ΔH_f^0 (σ-vinyl) $= 71\cdot1 \pm 2$ kcal mol^{-1} whence, using the heat of formation of vinylacetylene ($72\cdot8$ kcal mol^{-1}),[22] $ID(C_{sp}-C_{sp^2})$ turns out to be about 130 kcal mol^{-1}. The instantaneous bond dissociation energies are summarized in Table I.

So much for those cases where direct or indirect experimental evidence allows us to calculate, using conventional thermochemistry, instantaneous bond dissociation energies. Similar data for other bonds have to be obtained more indirectly, for example as follows. We will make use of the following abbreviations:

$$ID(C_{sp^3}-H) = a \qquad ID(C_{sp^2}=C_{sp^2}) = d$$
$$ID(C_{sp^3}-C_{sp^3}) = b \qquad \Delta H(C_{sp^3} \rightarrow C_{sp^2}) = e$$
$$ID(C_{sp^3}-C_{sp^2}) = c$$

The quantity e refers to the energy change on rehybridization of an sp^3 bonded carbon atom to sp^2. The bond dissociation of methane (12),

$$CH_4 \rightarrow CH_3{}^{\cdot} + H^{\cdot} \qquad \Delta H^{\circ} = 104 \pm 1 \text{ kcal mol}^{-1} \qquad (12)$$

where the methyl radical is produced in the stable planar ground state, can be built up in stages as follows:

(i). The instantaneous dissociation of a hydrogen atom to give the σ-methyl radical,

$$CH_4 \rightarrow \sigma\text{-}CH_3{}^{\cdot} + H^{\cdot} \qquad \Delta H^{\circ} = a \text{ kcal mol}^{-1} \qquad (13)$$

(ii). The instantaneous dissociation of three hydrogen atoms from the σ-methyl radical, produced in reaction (13),

$$\sigma\text{-}CH_3{}^{\cdot} \rightarrow C_{cp^3} + 3H^{\cdot} \qquad \Delta H^{\circ} = 3a \text{ kcal mol}^{-1} \qquad (14)$$

(iii). The rehybridization of the C_{sp^3} carbon atom,

$$C_{sp^3} \rightarrow C_{sp^2} \qquad \Delta H^{\circ} = e \text{ kcal mol}^{-1} \qquad (15)$$

(*iv*). The re-association of the three hydrogen atoms with the sp^2 hybridized carbon atom, exothermic to the extent of three times the C_{sp^2}—H instantaneous bond dissociation energy,

$$C_{sp^2} + 3H^\bullet \rightarrow \pi\text{-}CH_3^\bullet \qquad \Delta H^\circ = -333 \text{ kcal mol}^{-1} \qquad (16)$$

Using Hess's Law we find,

$$4a + e = 437 \pm 4 \text{ kcal mol}^{-1} \qquad (17)$$

Precisely the same treatment can be applied to the bond dissociation $D(C_2H_5—H)$, giving a π-ethyl radical and a hydrogen atom,

$$C_2H_6 \rightarrow \pi\text{-}C_2H_5^\bullet + H^\bullet \qquad \Delta H^\circ = 98 \pm 1 \text{ kcal mol}^{-1} \qquad (18)$$

to give,

$$98 = a - 2(111 - a) - (c - b) + e \qquad (19)$$

$$a + c - d = 117 \pm 3 \text{ kcal mol}^{-1} \qquad (20)$$

using the value for e derived in eqn. (17).

In order to solve the equations leading to values for the i.b.d.e.'s, it is necessary to introduce another quantity; in the original paper[1] this was derived from the value reported for the energy required to remove a hydrogen atom from the methyl group of propylene in a hot tritium recoil experiment.[26] The value observed, 93 kcal/mol^{-1} is greater than would be expected if the product was a ground state resonance stabilized allyl radical but less than that required to remove a hydrogen atom from methane and was supposed to reflect the inability of the resulting allyl radical to attain its most stable state. It was assumed by comparison and correlation with results obtained using other hydrocarbons that the allyl radical had attained the π-configuration but that conjugation had not been established. It is now thought[11] that the basic premise of this correlation, that the saturated radicals have sufficient time to relax to their ground state after hydrogen abstraction from the parent hydrocarbon by a hot tritium atom, may be false and that the comparisons ought to be with i.b.d.e.'s. It is suggested that this does not seriously affect the basic correlation and the calculations are as used previously:

$$CH_3CH{=}CH_2 \rightarrow \text{unconjugated } \pi\text{-}^\bullet CH_2CH{=}CH_2 \qquad (21)$$

$$\Delta H^\circ = 93 \pm 1 \text{ kcal mol}^{-1}$$

$$93 = a - 2(111 - a) - (116 - c) + e \qquad (22)$$

i.e. $\qquad a - c = 6 \pm 6 \text{ kcal mol}^{-1} \qquad (23)$

A value for d can be derived as follows:

$$\pi\text{-}CH_3CH_2^\bullet \rightarrow CH_2{=}CH_2 + H^\bullet \qquad (24)$$

$$\Delta H^\circ = 38{\cdot}7 \pm 1\,\text{kcal mol}^{-1}$$

$$38{\cdot}7 = a - 2(111 - a) - (d - c) + e \tag{25}$$

whence
$$d = 171{\cdot}3 \pm 7\,\text{kcal mol}^{-1} \tag{26}$$

Further,

$$C_2H_4 + H_2 \rightarrow C_2H_6 \tag{27}$$

$$\Delta H^\circ = -32{\cdot}8 \pm 1\,\text{kcal mol}^{-1}$$

$$32{\cdot}8 = 6a - 444 + b - 171 - 2e - 104{\cdot}2 \tag{28}$$

$$14a + b = 1626 \pm 7\,\text{kcal mol}^{-1} \tag{29}$$

We are now in a position to calculate all the unknown quantities from eqns. (20), (23) and (19) and we find that, $a = 109 \pm 2$, $b = 100 \pm 10$, $c = 103 \pm 6$ and $e = 0\,\text{kcal mol}^{-1}$. But it is possible to obtain a better approximation for c using the value of a to derive $\Delta H_f^\circ(\sigma\text{-}CH_3^{\cdot}) = 39 \pm 2\,\text{kcal mol}^{-1}$ and also the reaction,

$$CH_3CH{=}CH_2 \rightarrow \sigma\text{-}CH_3^{\cdot} + \sigma\text{-}C_2H_3^{\cdot} \tag{30}$$

whence,

$$c = 39 + 71 - 4{\cdot}9$$
$$= 105 \pm 4\,\text{kcal mol}^{-1}$$

A similar treatment on eqn. (31),

$$C_2H_6 \rightarrow 2\sigma\text{-}CH_3^{\cdot} \tag{31}$$

gives a first improved value for b of $98 \pm 3\,\text{kcal mol}^{-1}$.

We can now tabulate the calculated values of i.b.d.e.'s.

Table I
Instantaneous bond dissociation energies (i.b.d.e.'s)

Bond	i.b.d.e. (kcal mol^{-1})	Bond	i.b.d.e. (kcal mol^{-1})
C_p—H	81 ± 1	C_{sp2}—C_{sp2}	116 ± 3
C_{sp3}—H	109 ± 2	C_{sp2}—C_{sp}	130^a
C_{sp2}—H	111 ± 2	C_{sp}—C_{sp}	150^a
C_{sp}—H	129^a	$C_{sp2}{=}C_{sp2}$	171 ± 7
C_{sp3}—C_{sp3}	98 ± 3	$C_{sp2}{=}C_{sp}$	174 ± 6
C_{sp3}—C_{sp2}	105 ± 4	$C_{sp}{\equiv}C_{sp}$	211^a
C_{sp3}—C_{sp}	126^a		

^{a} Error limits not known.

The values of $ID(C_{sp2}{=}C_{sp})$ and $ID(C_{sp}{\equiv}C_{sp})$ were calculated from the heats of hydrogenation of allene and acetylene respectively.

A further interesting quantity which we are now in a position to calculate is that of the energy of valence state carbon above its ground state. As we have already noted, previous estimates of this[27] range upwards from $100 \, kcal \, mol^{-1}$, but using the above figures the heat of the reaction,

$$CH_4 \rightarrow C(\text{valence state}) + 4H^{\bullet}$$

$$\Delta H^{\circ} = 436 \pm 4 \, kcal \, mol^{-1}$$

exceeds that of the atomization of methane, $397 \cdot 2 \, kcal \, mol^{-1}$, by about $39 \pm 4 \, kcal \, mol^{-1}$ only, the valence state energy of carbon. The reason for this discrepancy is not clear.

It is also possible to use values of i.b.d.e's to calculate normal bond dissociation energies in hydrocarbons but in a number of cases consistent discrepancies were found between the values so calculated and those obtained from experimental methods. Among the simple hydrocarbons this effect was noticed when the hydrocarbon was branched; for example the calculated bond dissociation energies,

$$C_3H_8 \rightarrow iso\text{-}C_3H_7^{\bullet} + H^{\bullet}$$

$$iso\text{-}C_4H_{10} \rightarrow tert\text{-}C_4H_9^{\bullet} + H^{\bullet}$$

namely, $92 \cdot 6 \pm 6$ and $88 \pm 8 \, kcal \, mol^{-1}$ respectively appear lower than the experimental values, $94 \cdot 5 \pm 1$ and $91 \pm 1 \, kcal \, mol^{-1}$. In other words propane and isobutane seem rather more stable than expected.

HEATS OF ATOMIZATION

The same effect was noted when i.b.d.e's were used to calculate the heats of atomization of a number of alkanes and in order to account for the apparently greater stability of hydrocarbons greater than ethane it was found necessary to add to the heats of atomization, the structure parameters shown in Table II and to subtract the 1,4-gauche interaction energies shown in Table III. With these corrections and the values in Table Ia for the instantaneous bond dissociation energies, the agreement between experiment and theory (Table IV) is good. An increased number of interaction terms would no doubt give even better agreement in those cases where the estimated heats of atomization exceed those observed by considerably more than the error of their measurement[28] but such an extension to a simple scheme seems unwarrantable at present.

A number of alkenes, alkynes and diolefins were treated in the same way using the above structure parameters for the saturated part of the molecule and one of $1 \cdot 3 \, kcal \, mol^{-1}$ for the group $C_2C =$ (Tables V and VI).

Table Ia

Bond	i.b.d.e. (kcal mol^{-1})
C_{sp^3}—H	109
C_{sp^2}—H	111
C_{sp^3}—C_{sp^3}	97·6
C_{sp^3}—C_{sp^2}	105·3
C_{sp^2}—C_{sp^2}	116
C_{sp^2}=C_{sp^2}	171·3

Where appropriate the 1,4 interaction parameters across a double bond shown in Table VII were used.

Agreement between observed and calculated heats of atomization in the olefin series is as good as for the alkanes and the average deviation for the 13 olefins treated by Skinner and Pilcher,[29] of $\pm 0\cdot 2$ kcal mol^{-1}, compares well with those derived from empirical schemes and that of Dewar and Schmeising.[30] The latter approach bears some similarity to that used here in that hyperconjugation is not explicitly included as a property of olefinic hydrocarbons nor indeed is resonance stabilization found to be a significant property of buta-1,4-dienes.

From its experimental and observed heats of atomization the resonance energy of benzene was found to be 25·1 kcal mol^{-1} and the heats of atomization of a small number of aromatic hydrocarbons are also included in Table VI; the calculations ignored interaction energies which appeared to be significant in substituted benzenes.

TABLE II

Structure parameters (kcal mol^{-1})

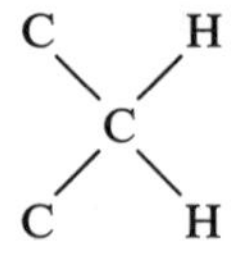

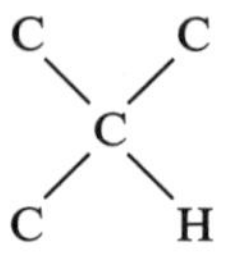

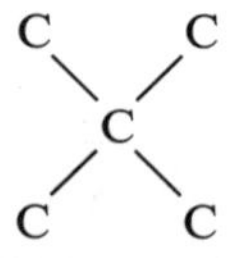

TABLE III

1,4-Interaction parameters (kcal mol^{-1})

C–CH$_2$ (with two C substituents on left C)	0·7	C–C (with two C substituents on each C)	1·0

TABLE IV

Errors of prediction of ΔH_a° *(25°, alkanes)*

Molecule	ΔH_a°(obs)[a,b]	Error[b]	Molecule	ΔH_a°(obs)[a,b]	Error[b]
C$_2$	674·6	+0·6	3-Et—C$_6$	2355·2	+0·1
C$_3$	954·3	−0·1	2,2-Me$_2$—C$_6$	2358·5	+0·1
C$_4$	1234·7	−0·3	2,3-Me$_2$—C$_6$	2356·0	+0·9
iso-C$_4$	1236·7	−0·1	2,4-Me$_2$—C$_6$	2357·3	+0·2
C$_5$	1514·6	0·0	2,5-Me$_2$—C$_6$	2358·0	+0·2
iso-C$_5$	1516·5	−0·4	3,3-Me$_2$—C$_6$	2357·4	−0·2
neo-C$_5$	1519·9[c]	−0·5	3,4-Me$_2$—C$_6$	2355·7	+0·5
C$_6$	1794·6	+0·2	3-Et-2-Me—C$_5$	2355·3	+0·9
2-Me—C$_5$	1796·3	0·0	3-Et-3-Me—C$_5$	2356·2	−0·4
3-Me—C$_5$	1795·7	−0·1	2,2,3-Me$_3$—C$_5$	2357·4	+0·1
2,2-Me$_2$—C$_4$	1799·0	−0·8	2,2,4-Me$_3$—C$_5$	2358·4	+1·7
2,3-Me$_2$—C$_4$	1797·2	0·0	2,3,3-Me$_3$—C$_5$	2356·6	+0·2
C$_7$	2074·6	+0·4	2,3,4-Me$_3$—C$_5$	2356·8	+1·0
2-Me—C$_6$	2076·3	+0·2	2,2,3,3-Me$_4$—C$_4$	2358·8	0·0
3-Me—C$_6$	2075·7	+0·1	C$_9$	2634·8	+0·8
3-Et—C$_5$	2075·1	0·0	3,3-Et$_2$—C$_5$	2634·6	−0·6
2,2-Me$_2$—C$_5$	2079·0	−0·6	2,2,3,3-Me$_4$—C$_5$	2636·6	+1·0
2,3-Me$_2$—C$_5$	2077·4	−0·7	2,2,3,4-Me$_4$—C$_5$	2638·8	−0·2
2,4-Me$_2$—C$_5$	2078·0	0·0	2,2,4,4-Me$_4$—C$_5$	2638·0	−3·2
3,3-Me$_2$—C$_5$	2077·9	−0·9	2,3,3,4-Me$_4$—C$_5$	2636·6	0·0
2,2,3-Me$_3$—C$_4$	2078·7	−0·7			
C$_8$	2354·6	+0·6	Average Deviation = ±0·5		
2-Me—C$_7$	2356·3	+0·3			
3-Me—C$_7$	2355·6	+0·3			
4-Me—C$_7$	2355·5	+0·3			

[a] API Project 44 Tables; [b] In kcal mol^{-1}; [c] Pilcher G. and Chadwick, J. D. M. (1967). *Trans. Faraday Soc.* **63**, 2357.

TABLE V

Errors of prediction of ΔH_a° (25°, mono-olefins)

Molecule	ΔH_a° (obs)[a,b]	Error[b]
C_2	537·7	0·0
C_3	820·3	−0·1
C_4-1	1100·4	0·0
cis-C_4-2	1101·8	−0·1
trans-C_4-2	1102·8	−0·1
iso-C_4	1104·0	0·0
C_5-1	1380·5	+0·1
cis-C_5-2	1382·2	−0·3
trans-C_5-2	1383·1	−0·3
2-Me—C_4-1	1384·2	0·0
3-Me—C_4-1	1382·4	−0·6
2-Me—C_4-2	1385·7	−0·2
C_6-1	1660·6	+0·2
cis-C_6-2	1662·2	−0·1
trans-C_6-2	1663·2	−0·1
cis-C_6-3	1662·2	−0·1
trans-C_6-3	1663·2	−0·1
2-Me—C_5-1	1664·2	+0·2
3-Me—C_5-1	1661·2	−0·2
4-Me—C_5-1	1662·3	0·0
2-Me—C_5-2	1665·6	+0·1
3-Me—cis-C_5-2	1664·9	+0·1
3-Me—trans-C_5-2	1664·9	+0·1
4-Me-cis-C_5-2	1663·9	−0·6
4-Me—trans-C_5-2	1664·9	−0·6
2-Et—C_4-1	1663·5	−0·1
2,3-Me_2—C_4-1	1665·4	+0·2
3,3-Me_2—C_4-1	1664·9	−0·3
2,3-Me_2—C_4-2	1666·5	+1·8
Average deviation		±0·2

[a] API Project 44 Tables; [b] kcal mol^{-1}.

STRAIN ENERGIES

The strain energies of a number of cyclic hydrocarbons calculated using the instantaneous bond dissociation energy method (Table VIII) are also in good agreement with those quoted by Skinner and Pilcher.[29]

FREE RADICALS

Table VI gives the resonance energies of a small number of hydrocarbon free radicals calculated by the present method. The value for allyl may

TABLE VI

Errors of prediction of ΔH_a° and estimated resonance energies (25°, dialkenes, allenes, alkynes, aromatic compounds and radicals)

Molecule	$\Delta H_a^\circ(\text{obs})^{a,b}$	Error[b]	Resonance energy
Diolefins			
C_4-1,3	969·9	−0·2	
cis-C_5-1,3	1252·6	−0·4	
trans-C_5-1,3	1252·5	−0·3	
2-Me—C_4-1,3	1253·2	+0·3	
C_5-1,4	1246·1	+0·5	
Allenes			
C_3-1,2	675·2	−0·2	
C_4-1,2	957·5	0·0	
3-Me—C_4-1,2	1240·3	−0·2	
C_5-1,2	1236·5	+1·2	
C_5-2,3	1238·2	+1·8	
Alkynes			
C_2	391·8	−0·4	
C_3	676·8	−0·2	
C_4-1	956·7	+0·1	
C_4-2	961·2	+0·6	
C_5-1	1236·8	+0·2	
C_5-2	1240·5	+1·5	
3-Me—C_4-1	1238·7	+0·5	
Aromatic compounds			
Benzene	1318·2	0·0	25·1
Toluene	1601·1	+1·6	25·1
Styrene	1748·8	+3·1	25·1
Biphenyl	2529·2	+5·0	50·2
Radicals			
Allyl			9·3
Cyclohexadienyl			19·7

[a] API Project 44 Tables; [b] kcal mol^{-1}.

be compared with that of Golden et al.[31] 10·2 kcal mol^{-1}, Golden, Gac and Benson,[32] 9·6 kcal mol^{-1}, Trenwith,[33] 12·4 ± 1·8 kcal mol^{-1} and Rogers et al.,[34] 9·4 kcal mol^{-1}. Small carbon radicals are treated in Table IX. The calculated heat of formation of C_pH is very close to the literature value[35] but the accepted values for $\Delta H_f^\circ(CH_2) = 91\cdot7 \pm 1$ kcal mol^{-1}[36] agree with that calculated here for the bent, excited, configuration. It may be that here again we have the case of a radical which can be formed and can react in an excited state on the input of sufficient energy, in this case during

TABLE VII

1,4-*Interaction parameters (kcal mol^{-1})*

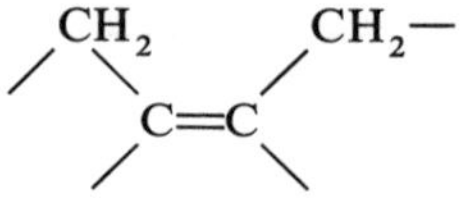

$$
\begin{array}{c}
CH_2 \qquad CH_2- \\
\diagup \qquad \diagup \\
\quad C=C \qquad\qquad 1{\cdot}0 \\
\diagup \qquad \diagdown \\
\end{array}
$$

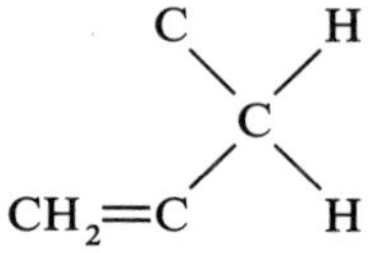

$$
\begin{array}{c}
C \qquad H \\
\diagdown \quad \diagup \\
\quad C \qquad\qquad 1{\cdot}0 \\
\diagup \quad \diagdown \\
CH_2=C \qquad H \\
\end{array}
$$

electron impact or photolysis.[36, 37] If the excited species attacks a substrate molecule or is otherwise detected before collisional or other deactivation to the linear ground state it will appear to be more endothermic than the ground state radical. The suggestion that two states of triplet methylene are involved in its reaction with methyl ethyl ether[38] is in line with this finding. On the other hand if the value of 91·7 kcal mol^{-1} for the heat of formation of methylene refers to the ground state of this species, then we must modify the i.b.d.e.'s of sp-hybridized carbon atoms as follows:

Bond	I.b.d.e. (kcal mol^{-1})
$C_{sp}-H$	112 ± 2
$C_{sp^3}-C_{sp}$	109 ± 2
$C_{sp^2}-C_{sp}$	113 ± 4
$C_{sp}-C_{sp}$	115 ± 4
$C_{sp}\equiv C_{sp}$	245 ± 4

but these values seem to be out of line with the others given in Table I.

STRUCTURE FACTORS

The structure factors clearly reflect the attractive interactions between the bonds or atoms in the molecule and it is of interest to speculate whether these are in some way an indication of the barriers to internal rotation in hydrocarbons, these being composed partly of energy required to break the attractive forces and partly of repulsive forces between the atoms or bonds in the eclipsed positions. The calculated size of the structure factors is not incompatible with this suggestion as can be seen in the comparison

TABLE VIII

Estimated strain energies

Molecule	$\Delta H_f^{\circ\,a,\,b}$	S_R^b
Cycloalkanes		
C_3	12·72	28·0
C_4	6·35	26·8
C_5	− 18·46	7·0
C_6	− 29·43	1·2
C_7	− 28·52	7·2
C_8	− 30·03	10·8
C_9	− 32·14	13·8
C_{10}	− 37·13	13·9
C_{11}	− 43·11	13·0
C_{12}	− 55·83	5·4
C_{13}	− 59·29	7·0
C_{14}	− 68·31	3·1
C_{15}	− 72·25	3·2
C_{16}	− 77·08	4·5
Cycloalkenes[c]		
C_3	66·6	64·4
C_4	35·0[d]	27·9
C_5	7·73	5·7
C_6	− 1·28	1·8
C_7	− 2·0	6·2
cis-C_8	− 7·0	6·3
trans-C_8	2·0	15·3
cis-C_9	− 8·5	9·9
trans-C_9	− 5·6	12·8
cis-C_{10}	− 16·3	7·0
trans-C_{10}	− 13·0	10·5
Cyclodialkenes[c]		
C_5-1,3	32·4[e]	5·9
C_6-1,3	26·0[f]	4·6
C_6-1,4	26·0[f]	1·6

[a] Ref. 31; [b] kcal mol^{-1}; [c] Interaction parameters across the double bond omitted; [d] Ref. 1; [e] Ref. 24; [f] James, D. G. L. and Suart, R. D. (1968). *Trans. Faraday Soc.* **64**, 2735.

between their value and those of the total rotational barriers in ethane, propane, isobutane and neopentane (Table X).[39] The absolute value for ethane does not agree since this effect was ignored in the calculations of the i.b.d.e.'s.

Although we have ignored the effect in ethane experiment has shown that it does exist, and that there is a barrier to rotation in ethane. Consequently, if the barrier is due to attraction between hydrogen atoms, removal of one

TABLE IX

$\Delta H°$ calculated by instantaneous bond dissociation energy
(kcal mol^{-1}) method

Reaction	$\Delta H°$
$\pi\text{-}CH_3^{\cdot} \rightarrow C_{sp^2}\text{—}H_2^{\cdot} + H^{\cdot}$	111 ± 2
$C_{sp^2}\text{—}H_2^{\cdot} \rightarrow C_{sp}\text{—}H_2^{\cdot}$	-36
$C_{sp}\text{—}H_2^{\cdot} \rightarrow C_{sp}H + H^{\cdot}$	129
$C_{sp}H \rightarrow C_pH$	9

whence

$$\Delta H_f°(C_{sp^2}H_2) = 93 \pm 3 \, \text{kcal mol}^{-1}$$

$$\Delta H_f°(C_{sp}H_2) = 57 \, \text{kcal mol}^{-1}$$

$$\Delta H_f°(C_pH) = 143 \, \text{kcal mol}^{-1}$$

TABLE X

Hydrocarbon	Structure factor (kcal mol^{-1})	Rotational energy barrier (kcal mol^{-1})
Ethane	0	2·9
Propane	3·4	6·8
Isobutane	9·0	10·8
Neopentane	15·0	17·2

or more of the hydrogen atoms from the ethane structure will necessitate some correction in the calculations. It is this that is suggested to be the reason for the necessity to subtract a small amount of energy for 1,4-interactions (Table IV).

One further shortcoming of this treatment is that it ignores the possible effects of hyperconjugation. In the thermodynamics of reaction (18), the dissociation of a hydrogen atom from ethane it is implicitly assumed that the reduction in the b.d.e. from the corresponding value for methane is due to the greater strength of the $C_{sp^2}\text{—}C_{sp^3}$ bond over that of $C_{sp^3}\text{—}C_{sp^3}$ compared to the difference in strength between $C_{sp^3}\text{—}H$ and $C_{sp^2}\text{—}H$. Unfortunately not enough is known of the energy effects of hyperconjugation for any meaningful correction to be made to the overall treatment although the results from e.s.r. spectroscopy have, as we have noted in Chapter 2, proved the existence of this phenomenon.

REFERENCES

1. Benson, S. W. J. (1965). *Chem. Ed.* **42**, 502.
2. Hartley, D. B. (1967). *Chem. Comm.* 1281.
3. Burton, C. S. and Noyes W. A. Jr. (1969). *In* "Comprehensive Chemical Kinetics", Vol. 3, 1.
4. Porter, R. P. and Noyes, W. A. Jr. (1959). *J. Amer. Chem. Soc.* **81**, 2307.
5. Terenin, A. and Werymin, H. (1935). *J. Chem. Phys.* **3**, 436.
6. McDowell, C. A. (1963). *In* "Mass Spectrometry", McGraw Hill, New York, 1963.
7. Batt, L. and Benson, S. W. (1962). *J. Chem. Phys.* **36**, 895; (1963). **38**, 3031.
8. Szwarc, M. (1950). *Chem. Rev.* **47**, 75.
9. Golden, D. M. and Benson, S. W. (1969). *Chem. Rev.* **69**, 125.
10. Breckenridge, W., Root, J. W. and Rowland, F. S. (1963). *J. Chem. Phys.* **39**, 2734; Root, J. W. and Rowland, F. S. (1964). *J. Phys. Chem.* **68**, 1226; Root, J. W., Breckenridge, W. and Rowland, F. S. (1965). *J. Chem. Phys.* **43**, 3694; Tachikawa, E. and Rowland, F. S. (1968). *J. Amer. Chem. Soc.* **90**, 4767.
11. Tachikawa, E. and Rowland, F. S. (1969). *J. Amer. Chem. Soc.* **91**, 559.
12. Evans, M. G. and Polanyi, M. (1938). *Trans. Faraday Soc.* **34**, 11.
13. Semenov, N. N. (1958). "Some Problems of Chemical Kinetics and Reactivity", Pergamon Press, London.
14. Kerr, J. A. (1966). *Chem. Rev.* **66**, 465.
15. Johnston, H. S. (1961). *Adv. Chem. Phys.* **3**, 131.
16. Cottrell, T. L. (1958). "The Strengths of Chemical Bonds", Butterworths, London.
17. Cochran, E. L., Adrian, F. J. and Bowers, V. A. (1964). *J. Chem. Phys.* **40**, 213.
18. Solly, R. K. and Benson, S. W. (1971). *J. Amer. Chem. Soc.* **93**, 1592.
19. Hay, J. M. (1970). *J. Chem. Soc. B*, 45.
20. Setzer, D. W. and Stedman, D. H. (1968). *J. Chem. Phys.* **49**, 467.
21. Cottrell, T. L. (1958). "The Strengths of Chemical Bonds", Butterworths, London.
22. Stull, D. R., Westrum, E. F. Jr. and Sinke, G. O. (1969). "The Chemical Thermodynamics of Organic Compounds", Wiley, New York.
23. Rodgers, A. S., Golden, D. M. and Benson, S. W. (1967). *J. Amer. Chem. Soc.* **89**, 4578.
24. Mortimer, C. T. (1962). "Reaction Heats and Bond Strengths", Pergamon Press, London.
25. Streitwieser, A. Jr. (1961). "Molecular Orbital Theory for Organic Chemists", Wiley, New York.
26. Tachikawa, E., Ni Yao Tang and Rowland, F. S. (1968). *J. Amer. Chem. Soc.* **90**, 3584.
27. Coulson, C. A. (1961). "Valence", 2nd Edn., Oxford University Press; Pilcher, G. and Skinner, H. A. (1962). *J. Inorg. Nucl. Chem.* **24**, 937.
28. Somayajulu, G. R. and Zwolinski, B. J. (1966). *Trans. Faraday Soc.* **62**, 2327.
29. Skinner, H. A. and Pilcher, G. (1963). *Quart. Rev.* **17**, 264.
30. Dewar, M. J. S. and Schmeising, H. N. (1959). *Tetrahedron* **5**, 166; (1960). **11**, 96.
31. Golden, D. M., Rodgers, A. S. and Benson, S. W. (1966). *J. Amer. Chem. Soc.* **88**, 3194, 3196.
32. Golden, D. M., Gac, N. A. and Benson, S. W. (1969). *J. Amer. Chem. Soc.* **91**, 2136.
33. Trenwith, A. B. (1970). *Trans. Faraday Soc.* **66**, 2805.
34. Rodgers, W. A., Wu, M. C. R. and Kuitu, L. (1972). *J. Phys. Chem.* **76**, 918.

35. Janaf Thermochemical Tables, N.B.S. 1965.
36. Chupka, W. A. and Lifshitz, C. (1968). *J. Chem. Phys.* **48**, 1109.
37. Bell, J. A. and Kistiakowsky, G. B. (1962). *J. Amer. Chem. Soc.* **84**, 3417.
38. Voisey, M. A. (1968). *Trans. Faraday Soc.* **64**, 3058.
39. Blades, E. and Kimball, G. E. (1950). *J. Chem. Phys.* **18**, 630.

CHAPTER 4

THE FORMATION AND REACTIVITY OF RADICALS: NON-POLAR EFFECTS

In this chapter we consider the non-polar features of the formation and reactivity of organic free radicals. In that all reactions involve polar effects to a greater or lesser extent such a distinction is not altogether satisfactory either in concept or in practice but by choosing reactions of simple hydrocarbon radicals such as methyl we can reduce complications due to polar effects to a minimum.

We will concentrate mainly on the reaction,

$$RR' \rightarrow R^{\boldsymbol{\cdot}} + R''$$ (1)

which produces free radicals and which is accompanied by the reverse reaction,

$$R^{\boldsymbol{\cdot}} + R'' \rightarrow RR'$$ (−1)

Up to this point we have considered hydrocarbon radicals as being capable of existing in one or other of two extreme forms, π or σ,

and we have mentioned the two extreme cases of dissociation to give radicals, that giving π-radicals being characterized by a lower A-factor and activation energy (E) than that giving σ-radicals. This approach predicts that with the application to the bond being broken on the one hand of the minimum amount of energy and on the other of a vast excess of energy,[110] the extreme forms of the kinetic parameters could be observed but under the sort of conditions normally employed in the laboratory, the actual mechanism involved is most likely to be a compromise and the structure of the dissociating parts of the molecule to lie somewhere between the two extreme forms of the radical, i.e. between the σ-form of the parent molecule and the π-form of (usually) the stable radical. In Arrhenius terms a balance is struck for the particular conditions chosen between the A- and E-factors which produces the optimum rate under these conditions.

87

This argument, with obvious modifications, for example in the case of radicals stable in the σ-configuration, applies to the dissociation into radicals of all organic molecules. It is, however, impossible to prove directly given the lack of experimental methods which can permit the observation of the shapes of molecules undergoing reaction. We therefore have to fall back on inference from circumstantial evidence.

Intuitively, and as a first shot,[1] it might be expected that some measure of the tendency of radicals produced in a dissociation reaction to the π-state would be given by the degree of mesomeric stabilization of the relevant radicals: the more the radicals are resonance stabilized the more likely do we expect to find them in the π-state and the lower do we expect to find A. Such a deduction is borne out in the results shown in Table I. Particularly significant are the figures referring to the production of phenyl and acyl which are σ in the ground state and those referring to oxy and imino radicals which are more stable in the π-state thanks to conjugation.

The wide variations in radical type produced in the reactions considered in Table I make fruitless any attempt to relate the figures in any more quantitative manner but a very rough indication of the trend is given in Fig. 1. The reactions involved have, as a common feature, the production of a methyl and another radical or molecule. For purposes of comparison any effects on the kinetic parameters due to the methyl radical produced are assumed common to all the reactions and the data are presented as a plot of the decadic logarithm of the experimentally observed A-factor against the delocalization energy of the radical or molecule. The Hückel parameters of these and other calculation are as given by Streitwieser[2] and methyl is treated as a heteroatom.

The straight line in Fig. 1 has the equation,

$$\log_{10}A = [17\cdot5 - 2(DE)] \pm 1$$

where A is expressed in units of s^{-1} and DE is the delocalization energy in units of β. The same relationship is not, of course, expected for all types of dissociation because of differences in the mesomeric stabilization of both product radicals and, where the fragments are bulky, steric effects.

Returning to Table I, it is noteworthy that although as we have shown in Chapter 2, alkoxy radicals show a strong tendency to achieve their stable π-state, the unimolecular decompositions of dimethylether and dimethyl peroxide are associated with high A-factors while, on the contrary, the decomposition of methylhydroperoxide has a particularly low A-factor. A previous commentator[3] on the kinetic parameters of the decompositions of peroxides has also noted that the A-factor for the dissociation of dimethylperoxide is unexpectedly high compared to those of the others of the homologous series. It is tentatively suggested that the reason for this apparent

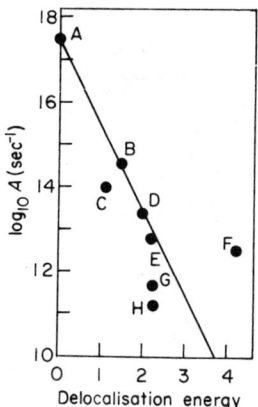

FIG. 1. Plot of $\log_{10} A$ for unimolecular decomposition giving methyl radicals against the delocalization energy of the products. A, CH_3-CH_3;[a] B, $C_6H_5CH_2-CH_3$;[b] C, CH_3-CH_2CN; [c] D, $C_6H_5NH-CH_3$;[d] E, $CH_3-CH(CH_3)CH_3$;[e] F, $CH_3-C(CH_3)_2CH_2$;[f] G, $CH_3CHCH_2-CH_3$;[g] H, $CH_3CHCH_2-CH_3$.[h]

[a] Benson, S. W. (1960). "The Foundations of Chemical Kinetics", McGraw-Hill, New York; [b] Esteban, G. L., Kerr, J. A. and Trotman-Dickenson, A. F. (1960). *J. Chem. Soc.* 3873; [c] Hunt, M., Kerr, J. A. and Trotman-Dickenson, A. F. (1965). *J. Chem. Soc.* 5074; [d] Esteban, G. L., Kerr, J. A. and Trotman-Dickenson, A. F. (1963). *J. Chem. Soc.* 3879; [e] Metcalfe, E. L. and Trotman-Dickenson, A. F. (1960). *J. Chem. Soc.* 5072; [f] Anderson, K. H. and Benson, S. W. (1964). *J. Chem. Phys.* **40**, 3747; [g] Gruver, J. T. and Calvert, J. G. (1956). *J. Amer. Chem. Soc.* **78**, 5208; [h] Bywater, S. and Steacie, E. W. R. (1951). *J. Chem. Phys.* **19**, 319.

inconsistency can be found in the compromise which we have discussed above. The marked tendency for the methoxy radicals to exist in the π-state leads to an A-factor for the decomposition of dimethylperoxide into ground state radicals which is so low that this process cannot compete with the alternative path into methoxy radicals with an essentially σ-configuration. A similar argument accounts for the very low A-factor for the decomposition of methylhydroperoxide and an extreme example is found in the decompositions of dimethyl- and tetramethylhydrazine which are not observable by the toluene carrier method.[4] Here again the resulting amino radicals are highly stable in the π-state and their formation must be accompanied by such low A-factors that the decompositions are undetectable in the toluene carrier experiment, the reaction conditions of which—relatively low temperatures, and in excess of toluene acting as an energy transfer agent—are such as to prevent high energy processes. The alternative decompositions into σ-radicals are therefore also so slow as to be unobservable. Table II gives an indication of the relative stabilities of methoxy and methylamino radicals, the bond energy data referring to the production of ground state radicals.

Figure 1 shows that we can explain the low A-factors of the dissociations of radicals using the same reasoning, since one of the products is an unsaturated

D

TABLE I

Arrhenius parameters of unimolecular decompositions

	$\log_{10} A\ [\sec^{-1}]$	$E\ (\text{kcal mol}^{-1})$	Ref.
$C_6H_5CH_2\!-\!COCH_3$	16·0	68·2	a
$CH_3CO\!-\!COCH_3$	15·7	66·0	a
$C_6H_5CO\!-\!Cl$	15·4	73·2	b
$C_6H_5CH_2\!-\!CO_2C_6H_5$	15·3	69·0	c
$CH_3\!-\!CH_3$	17·5	91·7	d
$(C_6H_5)_2Hg$	16·0	68·0	e
$CH_3O\!-\!CH_3$	17·5	81·0	f
$CH_3O\!-\!OCH_3$	15·4	36·1	g
$C_2H_5O\!-\!OC_2H_5$	13·32	31·7	h
$t\text{-}C_4H_9O\!-\!OC_4H_9\text{-}t$	15·6	37·4	i
$CH_3O\!-\!OH$	11 ± 2	32 ± 5	j
$CH_3NH\!-\!NH_2$	13·19	51·9	k
$C_6H_5NH\!-\!NH_2$	11·8	40·0	l

a Taylor, J. W. (1953). Ph.D. Thesis, University of Manchester; *b* Szwarc, M. and Taylor, J. W. (1954). *J. Chem. Phys.* **22**, 270; *c* Szwarc, M. and Taylor, J. W. (1953). *J. Chem. Phys.* **21**, 1746; *d* Quinn, C. P. (1963). *Proc. Roy. Soc. A*, **275**, 190; *e* Hartley, K., Pritchard, H. O. and Skinner, H. A. (1951). *Trans. Faraday Soc.* **47**, 254; *f* Anderson, K. H. and Benson, S. W. (1963). *J. Chem. Phys.* **39**, 1677; *g* Hanst, P. L. and Calvert, J. G. (1959). *J. Phys. Chem.* **63**, 103; *h* Rebbert, R. E. and Laidler, K. J. (1952). *J. Chem. Phys.* **20**, 574; *i* Batt, L. and Benson, S. W. (1962). *J. Chem. Phys.* **36**, 895; *j* Kirk, A. D. (1965). *Canad. J. Chem.* **43**, 2236; *k* Kerr, J. A., Sekhar, R. C. and Trotman-Dickenson, A. F. (1963). *J. Chem. Soc.* 3217; *l* Kerr, J. A., Trotman-Dickenson, A. F. and Wolter, M. (1964). *J. Chem. Soc.* 3584.

TABLE II

$$RXH \rightarrow RX^{\boldsymbol{\cdot}} + H^{\boldsymbol{\cdot}}$$

R	X	$D(RX\!-\!H)\ (\text{kcal mol}^{-1})$
H	O	118
CH_3	O	102
H	NH	103
CH_3	NH	92

molecule which is even more strongly stabilized in the π-state than a free radical. Consider the dissociation of the isopropoxy radical into acetaldehyde and methyl:

$$(CH_3)_2\underset{\displaystyle H}{C}O^{\boldsymbol{\cdot}} \rightarrow CH_3CHO + CH_3^{\boldsymbol{\cdot}}$$

with the following Arrhenius parameters,

$\log_{10} A$	$E(\text{kcal mol}^{-1})$
12·1	16·1[5]
11·8	17·2[6]

Theoretical analysis of these kinetic parameters leads to the conclusion that 24 and 20 oscillators respectively, in other words most of the 27 vibrational modes of the radical are involved in the reaction. This is entirely consistent with a picture of the dissociation which involves the production of both products in the π-state and during which extensive rehybridization involving the great majority of the bonds in the radical is taking place. Since the bonds are changing in length they are clearly involved in the distribution of energy throughout the radical. Even the methyl rotational modes may be involved since there is no strong evidence of hyperconjugation in the product aldehyde.

Other A-factors for alkoxy radical decomposition are:

$$CH_3CH_2O^{\bullet} \rightarrow CH_3^{\bullet} + CH_2O$$

$$\log_{10}A = 13\cdot4 \text{ (calculated)}[7]$$

$$= 11 \text{ (measured)}[7]$$

$$= 12\cdot15 \text{ (measured)}[8]$$

$$(CH_3)_3CO^{\bullet} \rightarrow CH_3COCH_3 + CH_3^{\bullet}$$

$$\log_{10} A = 9\cdot7 \pm 1\cdot2[7]$$

$$= 11\cdot2 \pm 1\cdot3[9]$$

$$= 16\cdot5 \text{ (calculated)}[10]$$

Lin and Laidler[11] have produced a relatively high A-factor for the decomposition of s-butyl:

$$s\text{-}C_4H_9^{\bullet} \xrightarrow{2} CH_3^{\bullet} + C_3H_6 \qquad k_2 = 6\cdot1 \times 10^{14} \exp(-32\,600/RT)$$

derived from the ratio,

$$k_2/k_3^{\frac{1}{2}} = 1\cdot31 \times 10^8\, e^{-32\,600}$$

in which k_3, the rate constant for the reaction,

$$2s\text{-}C_4H_9^{\bullet} \xrightarrow{3} C_8H_{18}$$

was assumed to be $2\cdot2 \times 10^{10}$ l.mol^{-1} s^{-1} and E_3 zero. The authors, however, suggest that A_3 and therefore A_2 may be somewhat lower. For the reactions,

$$1\text{-}C_3H_7^{\bullet} \rightarrow CH_3^{\bullet} + C_2H_4$$

$$2\text{-}C_6H_{13}^{\bullet} \rightarrow 1\text{-}C_3H_7^{\bullet} + C_3H_6$$

the values of $\log_{10} A$ were found to be $13\cdot6[12]$ and $13\cdot5[13]$ respectively.

The A-factor for the dissociation,

$$CH_3\dot{C}O \rightarrow CH_3^{\cdot} + CO$$

is extremely low, $10^{10.3}$ s^{-1}. O'Neal and Benson[14] explain this as being in some way the result of an adiabatic transition leading to a different electronic state of CO, but it may not be irrelevant to note that the π-bond energy of CO is very high and the tendency for it to attain the π-bonded state strong.

Dissociation involving the loss of a hydrogen atom would be expected to have a small polar component and also little steric hindrance so that we would expect such reactions to have a "normal" A-factor and this is in line with experiment.

It should also be noted that doubts as to the low A-factors found for radical decompositions have been expressed by Rabinowitch et al.[15] who preferred the following A-factors:

$$\log_{10} A$$

$$n\text{-}C_3H_7^{\cdot} \rightarrow CH_3^{\cdot} + C_2H_4 \qquad 15.36$$

$$s\text{-}C_4H_9^{\cdot} \rightarrow CH_3^{\cdot} + C_3H_6 \qquad 15.32$$
$$14.35$$

There has been some criticism of the author's presentation and that below dealing in a similar manner with metathetical reactions of free radicals both on account of the scatter of the experimental data and of their unreliability (two factors which may not be unrelated). It is undeniable that the method is unsophisticated but it is suggested that its significance lies in the fact that the relationships exist, i.e. their predictableness; no extensive claims are made for their predictiveness. Simply, application of the underlying principles enables the experimenter faced with an unexpected combination of A- and E-factors for a dissociation or metathesis involving simple radicals to decide whether these are reasonable; it would be irresponsible to claim that the stage has yet been reached, or perhaps ever will be reached, where the Arrhenius factors can be calculated with any degree of certainty.

Much the same sort of difficulty attaches to attempts which have been made to analyze in detail the kinetics of unimolecular decompositions and which require a knowledge of the disposition of the available energy throughout the molecule and product radicals during the process. The success of such calculations depends upon the number of degrees of freedom, both vibrational and rotational, permitted by the calculator to participate in the flow of energy throughout the molecule and radicals, there being no way to discover this experimentally, or indeed from *a priori* considerations unless one accepts a picture of the likely processes such as that presented above. The usual procedure, however, is merely to vary empirically the number of "effective" oscillators in the mathematical calculations without consideration of the chemical realities, until the "best" fit between experiment and theory is achieved. Perhaps the most conceptually satisfying development and the one which permits the least "fudging" is that of Marcus and Rice. This assumes that all the internal degrees of freedom of the molecule are active; the calculations however are somewhat complicated. As we have seen above a good illustration of a reaction in which the participation of all the degrees of freedom can easily be visualized is given by the decomposition of the iso-propoxy radical into a methyl radical and acetaldehyde.

Returning to our original set of reactions,

$$RR' \rightleftharpoons R^{\cdot} + R'' \tag{1}$$

the principle of microscopic reversibility assumes that the reaction path followed in the forward and reverse reactions is the same and in the case of the dissociation of a hydrocarbon into π-radicals it is easy to envisage the smooth rehybridization of the carbon atoms flanking the dissociating bond through an intermediate transitions state and, in the reverse reaction, the gradual distortion of the bonding around the carbon atom bearing the unpaired electron as the two radicals approach, through the same transition state and ultimately to the neutral molecule. Once the kinetics of a radical reaction are known and its thermodynamics measured or calculated, the principle of microscopic reversibility permits the calculation of the kinetics of the reverse reaction. In the case of reactions of type (1) above, the kinetics of the forward reaction are often known or readily measured and those of the reverse (-1) calculated using the assumption that the activation energy for the recombination of two free radicals is negligible. A large number of such systems have been treated in this way and when independent information has been obtained usually as the result of improvements in experimental technique of reaction study, the correlation between calculated and measured results has been found to be satisfactory.

But not always. Grotewold and his co-workers[16] have reported an elegant approach to the problem of observing radical recombination reactions, which as we have seen, having little or no energy barrier are extremely difficult to follow directly. The object of this work was to determine the pressure effect on methyl radical recombination rates using a competitive technique involving a comparison of the reactions,

$$R^{\cdot} + R^{\cdot} \rightarrow RR$$

$$R^{\cdot} + CH_3^{\cdot} \rightarrow RCH_3$$

$$CH_3^{\cdot} + CH_3^{\cdot} \rightarrow C_2H_6$$

If all three reactions were second order then changing pressure in the gas phase reaction system should have no effect on the relative concentrations of RR, RCH$_3$ and ethane formed. In a series of check reactions using ethyl and isopropyl as R$^\cdot$ it was shown that the ratios of their rates of disproportionation to combination, Δ(C$_2$H$_5^\cdot$, C$_2$H$_5^\cdot$) and Δ(i-C$_3$H$_7^\cdot$, i-C$_3$H$_7^\cdot$) were independent of pressure, in the former case down to 0·006 Torr. But in the competitive runs with methyl it was found that there was a pressure fall-off in the rate of methyl radical recombination some one hundred times lower than that obtained from calculation and experiment on the reverse reaction, the dissociation of ethane. From unimolecular rate theory and these recombination results the transition state for the recombination was calculated to have a much more rigid configuration than the results from the dissociation of ethane would suggest; s was calculated to be about 12·5. In a more recent paper[17] the recombination results were more nearly approximated using theoretical calculations based on a model with low frequency bending in the C—C bond, a much more rigid arrangement than one involving methyl groups free to rotate in three dimensions.[18]

This conclusion is easy to understand if it is assumed that the process of dissociation of ethane in the gas phase usually produces methyl radicals which are vibrationally excited and have an average structure which approximates to the σ-configuration. In other words, the compromise which we have discussed above in terms of A- and E-factors favours the production of radicals which tend to be in the σ-state. In the case of the reverse reaction, the recombination of methyl radicals, it is reasonable to assume that in the presence of other chemical species which can act as energy transfer agents, the methyl radicals coming together are in the planar or π-form and the transition state is more rigid. In terms of the pressure effect noted by the authors of ref. 16, the recombination of two planar methyl radicals involves an alteration in all the bonds in the incipient ethane molecule from sp^2 hybridization to sp^3; all six C—H bonds and the C—C bond are therefore able to participate in the dissipation of the energy of the C—C bond being formed and there is less tendency for the molecule to redissociate than there would be if the two radicals coming together were in the σ-state, in which case only the C—C bond would be readily available to take up the excess energy of the ethane molecule formed. The latter recombination therefore requires third body energy transfer and competes with redissociation at a higher pressure than does the former.

These effects do not show up in measurements of radical recombination rates obtained from data on the competing processes of recombination and hydrogen transfer.[19] In such experiments the competing processes,

$$R^\cdot + R^\cdot \xrightarrow{\;\;4\;\;} RR$$

$$R^\cdot + R'H \xrightarrow{\;\;5\;\;} RH + R''$$

are monitored by the rate of formation of RH. If the kinetics of reaction (5) are known from a separate experiment, the relationship,

$$d(RR)/dt = k_4(R^{\cdot})^2 = \frac{k_4 r_{RH}^2}{k_5^2(RH)^2}$$

gives us a measure of $(R^{\cdot})$ and hence k_4. Any increase in k_5 as the result of the methyl radical being in the σ-configuration will to some extent cancel out the consequent effect on k_4 and will therefore be undetectable.

Workers on the thermal decomposition of ethane, have however produced widely differing results for the Arrhenius parameters for the dissociation of ethane into two methyl radicals; these data, collected by Frey and Walsh[20] show a spread in A-values from $\log A = 16.0$ to 17.45 and in activation energies of $E = 86.0$ to 91.7 kcal mol^{-1}. And true to our picture of differing transition states under different energy conditions, Waage and Rabinovitch[21] have found it impossible to correlate this spread of results theoretically on the basis of a single model for dissociation and, by the principle of microscopic reversibility, of recombination. It is suggested that such efforts will prove fruitless except in the cases where the radicals are strongly stabilized in one or other of the two states, σ or π and that the best hope of correlating experiment with theory is to treat each set of data separately and to vary the configuration of the transition state bearing in mind the likely structures of the radicals involved.

Such an approach has been used in the combination reactions of methyl with isopropyl and tert-butyl radicals. Hase and Simons[22] found that using the entropy terms recommended for the π-radicals by O'Neal and Benson[23] and the assumption of a transition state with only slightly hindered methyl rotations, the calculated rate constants considerably underestimated those observed by experiment. It was found that a better correlation resulted from a transition state which was looser and in which the alkyl ends were tighter. This tightening was achieved by assuming a barrier to rotation of about 5 kcal mol^{-1}. By analogy with the partial structure,

$$\begin{array}{c} CH_3 \\ \diagdown \\ C{=} \\ \diagup \\ CH_3 \end{array}$$

it might be expected that the barrier to rotation in ground state alkyl radicals should be about 2 kcal mol^{-1} so that the structure used by Hase and Simons is closer to our picture of combination of two σ-radicals; we must assume that under the experimental conditions used by these authors the reacting radicals were vibrationally excited. See also more recent work.[24, 109]

Finally in this section it is not unreasonable to expect that dissociations involving the production of radicals stable in the π-state should be accompanied by low A-factors. And from the little evidence which is available, this appears to be the case (Table III).

TABLE III

		log A	E	Ref.
(structure)	$\rightarrow CH_3^{\cdot} + C_3H_5^{\cdot}$	9·7	59·1	a
		10·9	69·5	b
(structure)	$\rightarrow t\text{-Bu}^{\cdot} + C_3H_5^{\cdot}$	12·8	65·5	c
(structure)	$\rightarrow 2C_3H_5^{\cdot}$	10·4	56·0	d
		10·3	45·6	e
$C_6H_5CH_2CH_3$	$\rightarrow C_6H_5CH_2^{\cdot} + CH_3^{\cdot}$	11·6	70·1	f

[a] Kerr, J. A., Spencer, R. and Trotman-Dickenson, A. F. (1965). *J. Chem. Soc.* 6652; [b] Halstead, M. P. and Quinn, C. P. (1968). *Trans. Faraday Soc.* **64**, 103; [c] Tsang, W. (1967). *J. Chem. Phys.* **46**, 2817; [d] Akers, R. J. and Throssell, J. J. (1967). *Trans. Faraday Soc.* **63**, 124; Throssell, J. J. (1972). *Inter J. Chem Kinetics* **4**, 273; [e] Homer, J. B. and Lossing, F. P. (1966). *Canad. J. Chem.* **44**, 2211; [f] Esteban, G. L., Kerr, J. A. and Trotman-Dickenson, A. F. (1963). *J. Chem. Soc.* 3873.

RADICAL–RADICAL REACTIONS

Similar considerations are expected to apply to the reverse process of unimolecular dissociation, namely radical association and also radical disproportionation,

$$2RCH_2CH_2^{\cdot} \xrightarrow{k_d} RCH{=}CH_2 + RCH_2CH_3$$
$$\xrightarrow{k_c} (RCH_2CH_2)_2$$

$$2RCH_2O^{\cdot} \xrightarrow{k_d} RCH_2OH + RCHO$$
$$\xrightarrow{k_c} (RCH_2O)_2$$

$$2RCH_2NH^{\cdot} \xrightarrow{k_d} RCH{=}NH + RCH_2NH_2$$
$$\xrightarrow{k_c} (RCH_2NH)_2$$

Kinetic parameters for radical combination reactions are often derived from those of molecular dissociation and those for disproportionation from a knowledge of $\Delta = k_d/k_c$, which is often not very different from unity. Since, as we have suggested, dissociation reactions of hydrocarbons may be accompanied by relatively high A-factors due to the formation of the product radicals in the σ-state, or at least vibrationally excited to some extent, this method of measuring kinetic parameters of radical–radical reactions may overestimate the A-factors of reactions involving ground state π-radicals.

It will be instructive to consider in a purely pictorial manner radical–radical reactions in which the geometry of the radicals is at least moderately well understood and to attempt to relate the conclusions drawn to results from similar reactions where the conditions are not so well defined. Again, for simplicity of discussion, we will concentrate on examples where reaction occurs between radicals in the extreme structures, σ- or π.

(1) $\sigma + \sigma$

$$2RCH_2CH_2 \, \langle \cdot \rangle \rightarrow ?$$

Here the efficiency of overlap of the σ-orbitals is expected to be high and not to vary critically in direction over a wide angle giving rise to a high A-factor for combination and a low or zero activation energy. Since however the configurations of the two radicals do not change on combination, in the absence of a fortuitous resonance energy transfer between the bond being formed and the rest of the incipient molecule there is a good chance that the molecule will fly apart immediately on formation. Third body energy transfer would also assist combination. If some mechanism is available to dissipate the excess energy then it would be expected that overlap of the two σ-orbitals would produce more combination than disproportionation. There is no direct confirmation of this conclusion but it has been found that butyl radicals produced under high intensity radiation conditions which are likely to give vibrationally excited radicals, show a greater tendency to combine relative to disproportionation than do the ground state radicals.[25, 26] Another interesting example which we have already touched upon briefly is that of the combination of trichloromethyl radicals. In the gas phase these radicals when produced by photolysis of trichlorobromomethane combine with a rate constant, $k = 10^{10 \cdot 5}$[27] while in solution, where they are unlikely to be excited, $k = 10^{8}$.[28] An intermediate value of $10^{8 \cdot 8}$ has also been obtained in gas phase work.[29]

It is difficult to see how disproportionation could occur from the mutual interaction of half-filled σ-orbitals but a simple hydrogen abstraction could occur, such as,

$$X{\equiv}C\langle \cdot \rangle + H{-}\overset{\displaystyle Y}{\underset{\displaystyle Z}{C}}\langle \cdot \rangle \rightarrow X{\equiv}CH + C{\equiv}Y \ \overset{}{\underset{\displaystyle Z}{}}$$

for example,

$$CH_2{=}CH^{\cdot} + CH_2{=}CH^{\cdot} \rightarrow C_2H_4 + C_2H_2$$

But evidence from the low pressure hydrogenation of acetylene rather supports the reaction,

$$2C_2H_3{}^{\cdot} \rightarrow 2C_2H_2 + H_2$$

which could proceed via the energetically favoured six-membered ring arrangement,

Small cyclic radicals also tend to have a σ-configuration and in conformity with the above discussion, cyclopentyl radicals are found[30] to have $\Delta = 0.2$ which is less than that for the analogous non-cyclic hydrocarbon radical, iso-propyl, where Δ ranges from 0.5 to 2.

(2) σ + π

$$RCH_2CH_2 \;\; \oplus + \;\; \overset{\cdot}{C}H_2CH_2R \rightarrow ?$$

This arrangement leads to recombination with less likelihood of redissociation than that above since one of the radicals in changing hydridization on combination provides a possible mechanism for the dissipation of the energy of the bond being formed. Disproportionation giving head to tail abstraction as demanded by the experimental results[31] can also occur either by way of a four-membered ring transition state or by a direct hydrogen abstraction reaction,

From this simple picture it might be expected that the reaction between a σ and a π-radical would provide an efficient arrangement for disproportionation, a conclusion which is not at variance with the following results compiled by Cadman et al.[32] The greater the degree of halogen substitution of the radical the greater the σ-character and the greater the degree of disproportionation. The values in the table are all higher than the corresponding values for hydrocarbon radicals (see Table V) although in some cases the numbers of attackable hydrogen atoms are less.

TABLE IV

R˙, R″˙	Δ
$CF_3\dot{}$, $i\text{-}C_3H_7\dot{}$	$2\cdot2 \pm 0\cdot1^a$
$C_3F_7\dot{}$, $C_2H_5\dot{}$	$0\cdot40,^a\ 0\cdot2^b$
$C_2F_5\dot{}$, $C_2H_5\dot{}$	$0\cdot56^b$
$CCl_3\dot{}$, $C_2H_5\dot{}$	$0\cdot24,^c\ 0\cdot23^d$
$CHCl_2\dot{}$, $C_2H_5\dot{}$	$0\cdot07^d$
$CH_3CHF\dot{}$, $CH_3CHF\dot{}$	$0\cdot21^e$
$CH_3CF_2\dot{}$, $CH_3CH_2\dot{}$	$0\cdot55^e$

[a] Giacometti, G. and Steacie, E. W. R. (1958). *Canad. J. Chem.* **36**, 1493; [b] Pritchard, G. O. and Thommarson R. L., (1961). *J. Phys. Chem.* **70**, 3339, 2307; [c] Gregory, J. E. and Wijnen, M. H. J. (1963). *J. Chem. Phys.* **38**, 2925; [d] Yu, W. H. S. and Wijnen, M. H. J. (1970). *J. Chem. Phys.* **52**, 2736; [e] Scott, P. M. and Jennings, K. R. (1969). *J. Phys. Chem.* **73**, 1521.

A similar sort of argument can be employed to explain the reaction of methyl radicals with oxygen. The association reaction of methyl with oxygen is third order:

$$CH_3\dot{} + O_2 \xrightarrow{M} CH_3O_2\dot{}$$

but at low pressures a bimolecular reaction is observed which has been suggested to be followed by the rearrangement and rupture of the initially formed and excited methylperoxy radical to give formaldehyde and a hydroxy radical:[33]

$$CH_3O_2\dot{}* \rightarrow HCHO + \dot{}OH$$

a suggestion which has been criticized on the grounds that it incidentally involves the formation of a four-membered and therefore highly strained ring,[34]

Such an objection notwithstanding flash photolysis of methyl iodide–oxygen mixtures[35] and mass spectrometric analysis of the products of photolysis of methyl–iodide–oxygen mixtures[36] have shown that formaldehyde is a major initial product of the oxidation of methyl. The experimental methods employed in these studies involved the use of high energy photolytic sources which would tend to produce excited or σ-methyl radicals which can react with oxygen in a manner formally analogous to that discussed above for disproportionation,

$$HCHO + {}^{\bullet}OH$$

This explanation is consistent with results using flash photolysis.[106]

(3) π + π

This is the arrangement for most radical–radical reactions where the reactants are in their ground states and, from our arguments above on the reverse reactions, it is expected to give combination and/or disproportionation with a low *A*-factor. Alkoxy radical–alkoxy radical reactions provide a good example of this category,

$$RCH_2OOCH_2R'$$

$$RCHO + R'CH_2OH$$

The mechanism for disproportionation involves a four-membered ring transition state and is thus equivalent to that proposed by Bradley.[37] Criticisms of this mechanism have hinged on the likely magnitudes of *A*-factors for disproportionation. The argument goes that since Δ is usually not far from unity then *A*-factors for disproportionation must be similar to those for combination: *A*-factors for combination calculated from the reverse process of dissociation using the principle of microscopic reversibility, are high. Therefore *A*-factors for disproportionation must also be high. Transition states involving four-membered ring transition states must be highly strained and therefore have a low *A*-factor so that they cannot be intermediate in the process of disproportionation. But, as we have suggested above, *A*-factors for combination of radicals in the π-ground state may not after all be high: it follows that radical–radical reactions of disproportionation may also be accompanied by low A-factors. Proof of this argument one way of the other awaits direct experimental evidence on the kinetics of radical–radical reactions, but one indication is given by figures on the combination reaction of methyl and benzyl[38] where $\log k = 8.2$. Using the geometric rule,

$$k_{AB} = 2(k_{AA}k_{BB})^{\frac{1}{2}}$$

where the rate constants refer respectively to the reactions,

$$A + B \rightarrow AB$$
$$2A \rightarrow A_2$$
$$2B \rightarrow B_2$$

a very low rate constant of $10^{5 \cdot 5}$ l. mol^{-1} s^{-1} is derived for the combination of two benzyl radicals. Since the activation energy of this reaction, like that of other radical–radical reactions, is assumed to be zero or at least very low, it is inferred that A is also very low. Benzyl, of course, is highly stabilized as a π-radical.

Another interesting observation suggesting that the ratio of combination to disproportionation may be a function of the shape of the reactant radicals comes from work using photochemically produced radical pairs namely ethyl + isopropyl[39] and n-butyl + n-butyl.[40] Unlike most systems it was found that Δ increased with temperature. It may be suggested that the radicals in these systems were formed initially in an excited and therefore σ-state; increase in temperature helped prolong the lifetimes of these excited species before deactivation so that the number of radical–radical encounters leading to reaction which involved a σ-species increased as the temperature was raised. It is not likely that reactions involving σ-radicals would exhibit the same value of Δ as those between π-state radicals. This explanation also would account for the absence of a temperature effect in the reactions of thermally stable radicals since it is highly unlikely that the mere increase of temperature could preferentially excite the vibration modes leading to the production of σ-radicals.

Table V sets out the values found for Δ for a series of hydrocarbon and a number of alkoxy radicals. Because of the different experimental conditions used by the authors and the difficulty of predicting the configuration of the reacting radicals it is scarcely to be expected that any very obvious trend will emerge from the figures; it is however, noteworthy that in the case of alkoxy radicals which are more highly stabilized in the π-state than the corresponding hydrocarbon radicals, the ratio k_d/k_c very roughly follows the numbers of abstractable hydrogen atoms. For alkyl radicals Terry and

Futrell[41] have derived the empirical relationship,

$$\log k_d/k_c = 0\cdot111(S_d^\circ - S_c^\circ) - 4\cdot88$$

where S° represents the standard entropy of the relevant products. The equation accords well with the results from methyl and ethyl radicals but not so satisfactorily with 2-propyl and t-butyl.

TABLE V

R˙, R′˙	Δ
$CH_3\cdot$, $C_2H_5\cdot$	0·036[a]
$CH_3\cdot$, iso-$C_3H_7\cdot$	0·21[b]
$CH_3\cdot$, n-$C_4H_9\cdot$	0·15[c]
$CH_3\cdot$, cyclo-$C_5H_9\cdot$	0·3[d]
$C_2H_5\cdot$, $C_2H_5\cdot$	0·14[b, l]
$C_2H_5\cdot$, iso-$C_3H_7\cdot$	0·43[b]
$C_3H_7\cdot$, $C_3H_7\cdot$	0·16,[e] 0·15[l]
n-$C_3H_7\cdot$, iso-$C_3H_7\cdot$	0·14[l]
iso-$C_3H_7\cdot$, iso-$C_3H_7\cdot$	0·63,[b] 0·69[l]
n-$C_4H_9\cdot$, n-$C_4H_9\cdot$	0·94,[c] 0·14[l]
cyclo-$C_5H_9\cdot$, cyclo-$C_5H_9\cdot$	1·0[f]
$CH_3\cdot$, $CH_3O\cdot$	1·6[g]
$CH_3\cdot$, iso-$C_3H_7O\cdot$	3·4[h]
$C_2H_5\cdot$, $C_2H_5O\cdot$	1·3[i]
$C_2H_5O\cdot$, $C_2H_5\cdot$	2·3[i]
$CH_3O\cdot$, $CH_3O\cdot$	9·3[j]
$C_2H_5O\cdot$, $C_2H_5O\cdot$	12[j]
iso-$C_3H_7O\cdot$, iso-$C_3H_7O\cdot$	0·61[k]

[a] Grotewold, J., Lissi, E. A. and Neumann, M. G. (1968). J. Chem. Soc. A, 375; [b] Thynne, J. C. J. (1961). Proc. Chem. Soc. 68; [c] Thynne, J. C. J. (1961). Proc. Chem. Soc. 18; [d] Gordon, A. S. (1965). Canad. J. Chem. 43, 570; [e] Papic, M. M., and Laidler, K. J. (1971). Canad. J. Chem. 49, 454; [f] Gunning, H. E. and Stock, R. L. (1964). Canad. J. Chem. 42, 357; [g] Yee Quee, M. J. and Thynne, J. C. J. (1966). Trans. Faraday Soc. 62, 3154; [h] McMillan, G. R. (1960). J. Amer. Chem. Soc. 82, 2422; [i] Wijnen, M. H. J. (1960). J. Amer. Chem. Soc. 82, 3034; [j] Heicklen, J. and Johnston, H. S. (1962). J. Amer. Chem. Soc. 84, 4030; [k] Cadman, P., Ivel, Y. and Trotman-Dickenson, A. F. (1970). J. Chem. Soc. A, 1207; [l] Falconer, W. E. and Surdeo, W. A. (1971). Int. J. Chem. Kinetics 3, 523.

RADICALS + O_2 AND NO

In this section it is convenient also to consider the reactions of free radicals with the stable π-free radicals, oxygen and nitric oxide. The measured combination rates for π-alkyl radicals with oxygen are around 10^8 l. mol^{-1} s^{-1} and are characteristic of what we would expect of combination of π-type radicals. In a typical hydrocarbon oxidation reaction in the gas phase or in solution the addition reaction between a radical and oxygen is followed by abstraction to give a peroxide,

$$R^{\cdot} + O_2 \rightarrow RO_2^{\cdot}$$
$$R'H + RO_2^{\cdot} \rightarrow R'' + ROOH$$

and the further behaviour of the reaction is usually controlled by the reactions of the hydroperoxide. The same process can occur with aldehydes, for example the oxidation of acetaldehyde at low temperatures produces peracetic and acetic acids, roughly as follows,

$$CH_3\dot{C}O + O_2 \rightarrow CH_3CO_3^{\cdot}$$
$$CH_3CO_3^{\cdot} + CH_3CHO \rightarrow CH_3CO_3H + CH_3\dot{C}O$$
$$\downarrow$$
$$CH_3COOH + products.$$

It has been suggested[42] that the disproportionation reaction,

$$\pi\text{-}RCH_2CH_2^{\cdot} + O_2 \rightarrow RCH{=}CH_2 + HO_2^{\cdot}$$

also occurs and is important in the slow combustion of hydrocarbons.

The reactions between σ-radicals and oxygen have been noted by the author;[43] it seems to be characteristic of these processes that the addition reaction is inefficient. Phenyl for example reacts with oxygen some six hundred times slower than do alkyl radicals[44] and even at 77°K where addition took place readily between π-radicals and oxygen there was no evidence of such a process between phenyl and oxygen.[45] Similarly, formyl does not readily add oxygen,[46] a fact which no doubt explains the low yields of formic and performic acids from its oxidation[47] and the difference between the characteristics of its oxidation (and that of glyoxal which is very similar),[48] and that of the oxidations of most other organic fuels. The reaction between formyl and oxygen is predominantly one of disproportionation,

$$^{\cdot}CHO + O_2 \rightarrow CO + HO_2^{\cdot}$$

Evidence relating to the reaction of vinyl with oxygen comes from work on the thermal chlorination of ethylene.[49] At temperatures below 285°C the chlorination proceeds via addition to give chloromethyl radicals. The process is inhibited by the presence of oxygen presumably via the addition of the π-chloroethyl radicals to the oxygen,

$$C_2H_4Cl^{\cdot} + O_2 \rightarrow C_2H_4ClO_2^{\cdot} \rightarrow chain\ end$$

Above 285°C the main part of the reaction is one of substitution,

$$C_2H_4 + Cl^{\cdot} \rightarrow HCl + C_2H_3^{\cdot} \rightarrow etc.$$

Under these conditions oxygen accelerates the reaction; the vinyl radical being σ does not readily add oxygen which we must assume merely acts as a radical initiator. Addition of ethane at these higher temperatures once again renders the reaction susceptibly to inhibition by oxygen, no doubt by the efficient removal by oxygen of the π-ethyl radical in the following sequence,

$$C_2H_6 + Cl^{\cdot} \text{ (or vinyl)} \rightarrow C_2H_5^{\cdot} + HCl \text{ (or ethylene)}$$

$$C_2H_5^{\cdot} + O_2 \rightarrow C_2H_5O_2^{\cdot} \rightarrow \text{chain end}$$

the ethylperoxy radical being much less reactive than either ethyl or vinyl.

An addition reaction between a vinyl radical formed by the addition of hydroxyl radicals to acetylene, and oxygen has been proposed to account for results from the slow combustion of acetylene[50] but the same radical is also suggested to be sufficiently reactive to abstract a hydrogen atom from acetylene,

where RH can be acetylene. The radical produced by oxygen addition to hydroxy vinyl is supposed to be able to undergo intra or intermolecular reactions,

$$(HOCH{=}CHO_2^{\cdot})^* + C_2H_2 \rightarrow HOCH{=}CHO_2^{\cdot} \xrightarrow{O_2} (CHO)_2 + {\cdot}OH$$

acetylene apparently acting as a specific third body in the latter reaction.

The reactions of nitric oxide with radicals are analogous,[51]

$$\cdot CHO + NO \rightarrow HNO + CO$$

$$\rightarrow \quad \begin{matrix} H \\ \diagdown \\ \quad C=O \rightarrow HCN + O_2 \\ \diagup \\ N \\ \diagdown \\ O\cdot \end{matrix}$$

$$\begin{matrix} H \quad\quad H \\ \diagdown \quad \diagup \\ C=C \quad \rightarrow HCHO + HCN \\ \diagup \quad\quad \diagdown \\ N \quad\quad H \\ \diagdown \\ O\cdot \end{matrix}$$

the excess energy in the C—N bond formed overcoming any activation energy required to form the small rings.

Oxidation of the σ-cyclopropyl radical at 380°–430°C is suggested[52] to involve a ring enlarging process,

no doubt again facilitated by the excess energy in the peroxy radical. No ring opening occurs with the phenyl radical although the following possible fate has suggested[53] for an initially formed phenyl peroxy radical,

There seems little doubt that the apparent reluctance of σ-radicals to combine with oxygen is explicable in terms of energy transfer efficiency.

As in the case of the mutual interaction of two σ-radicals, the association of a σ-radical with oxygen involves no rehybridization of the organic moiety, although the O—O bond may change sufficiently in the formation of a peroxy radical to absorb some of the excess energy. Unless the excited peroxy radical is deactivated by a third body collision or by an intra- or intermolecular reaction, it will redissociate and it will appear that σ-radicals do not react with oxygen.

It is natural to enquire at this point whether any of the conclusions we have drawn about the reactions of organic radicals with oxygen and nitric oxide can help in elucidating any of the gross features of the reactions of organic compounds with these species.

To take the former first, there are three highly characteristic features of the gas-phase oxidation of hydrocarbons and certain other organic fuels which are observed with most fuels in the temperature region of so-called slow combustion, i.e. somewhere between 200° and 500°C. The first is degenerate branching or the slow acceleration of the oxidation over a period of minutes or longer in contrast to the conventional free radical branching reaction which is usually measured in seconds or less. The second is the period of negative temperature coefficient in which the rate of the reaction decreases with increase in temperature before finally increasing again; and the third which is usually closely associated with the second is the phenomenon of cool flames. Cool flames are sudden accelerations of the combustion, accompanied by the emission of luminescence due to excited formaldehyde. The accelerations are, however, short lived and after quenching the reaction appears to proceed almost as though they had never occurred, although detailed chemical analysis of the products of reaction before, during and just after a cool flame shows that quite profound changes have taken place at the microscopic level. For details of these and other characteristics of the combustion of organic fuels the reader is referred to one or other of the standard works in the field.[54]

To take these phenomena in turn, degenerate branching is simply explained as a relatively slow multiplication of the radical stock of the reaction by a molecular, rather than a wholly radical process, such as occurs for example in the oxidation of hydrogen. The simplest molecular process is the decomposition of a peroxide,

$$ROOH \rightarrow RO^{\cdot} + {}^{\cdot}OH$$

or

$$RCOOOH \rightarrow RCO_2^{\cdot} + {}^{\cdot}OH$$

and the energy requirements of most slow combustions are compatible with such a mechanism. In the case of formaldehyde, however, and perhaps glyoxal, while the multiplication of radicals is probably also the result of peracid decomposition,

$$HCOOOH \rightarrow HCO_2^{\bullet} + {}^{\bullet}OH$$

the rate of branching which in most combustions is controlled by the rate of peroxide decomposition, seems to be controlled by the rate of formation of the peroxide. This is not really surprising in view of our discussion above since the excited and highly unstable peroxy radical, HCO_3^* formed by the association of the σ-formyl radical and oxygen will tend to redissociate and the formation of the peracid must compete with this preferred process. It has therefore been suggested[47] that the branching of formaldehyde oxidation is controlled by the radical process,

$$(HCO_3^{\bullet})^* + HCHO \rightarrow HCO_3H + {}^{\bullet}CHO$$

And the activation energy of the overall process is compatible with rate control by a radical process being around $17\,kcal\,mol^{-1}$ in contrast with values of upwards of $30\,kcal\,mol^{-1}$ for most organic fuel oxidations.

The occurrence of a region of negative temperature coefficient suggests a change-over in reaction mechanism accompanied by the suppression of a key low temperature mechanism and this too can be explained in terms of the above discussion if it is accepted, as suggested by Senenov[55] and shown by the author for acetylenes[56] that the overall branching mechanism in the oxidation of a hydrocarbon involves the formation and dissociation of a peracid,

$$RH \xrightarrow{\text{oxidation}} R'CHO$$

$$R'CHO + X^{\bullet} \longrightarrow R'CO^{\bullet} + XH$$

$$R'CO^{\bullet} + O_2 \rightleftharpoons (R'CO_3^{\bullet})^*$$

$$(R'CO_3^{\bullet})^* \xrightarrow{M} R'CO_3^{\bullet}$$

$$R'CO_3^{\bullet} + RH \longrightarrow R'CO_3H + R^{\bullet}$$

$$R'CO_3H \longrightarrow R'CO_2^{\bullet} + {}^{\bullet}OH$$

$$\longrightarrow \text{branching}$$

At low temperatures the carbonyl radical is more likely to be in a π-state, energy transfer will also tend to be more efficient, and there will be a sufficient concentration of peracyl radicals to produce the peracid for branching. As the temperature is raised, however, σ-character of the radicals becomes more marked, they tend to dissociate and/or abstract rather than add oxygen, the concentration of peracid drops and with it the branching rate and the rate of the reaction. This phenomenon of negative temperature coefficient is also exhibited by aldehydes and ethers which produce aldehydes on combustion.

Many suggestions have been put forward to explain the third phenomenon, that of cool flames, all of which must have one thing in common, namely a **chemical reaction exothermic to the extent of at least 77 kcal mol^{-1} and having formaldehyde as one of its products.** No-one can be sure as to the exact mechanism for the formation of the excited formaldehyde but the quantum yield of its formation at around 10^{-6} is very low. It is suggested therefore that it is the result of radical–radical processes and since the phenomenon is closely allied to the region of negative temperature coefficient, that the radicals involved are partly from the low temperature regime and partly from the high temperature régime, for example

$$RCH_2CO_3^{\cdot} + HO_2^{\cdot} \rightarrow ROH + HCHO + O_2 + CO$$

$$\Delta H = c. -80 \, \text{kcal mol}^{-1}$$

or

$$RCH_2CO_3^{\cdot} + \,^{\cdot}OH \rightarrow ROH + HCHO + CO_2$$

$$\Delta H = c. -150 \, \text{kcal mol}^{-1}$$

the actual mechanism proceeding via ring transition state,

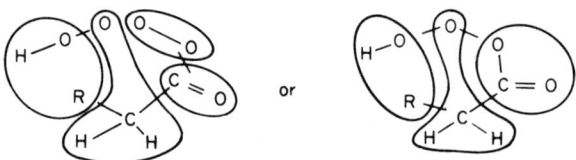

This suggestion is similar to that of Lewis and von Elbe,[57] and it also accounts for the resistance of branched chain hydrocarbons to knock and cool flame production[58] since these fuels do not have the necessary —CH$_2$— structure, α to the carbonyl group.

The reactions of organic molecules with nitric oxide are little less complicated that those involving oxygen but if it is assumed that basically the reactions of organic radicals with nitric oxide are similar to those with oxygen, i.e. readily reversible unless the resulting adduct is stabilized by efficient energy transfer, either internal by hybridization changes or accidental resonances or external by third body collision, then we have a possible explanation of the effects of adding nitric oxide to certain thermal decomposition reactions which produce radicals.

Consider first the series of carbonyl compounds in Table VI. We have

TABLE VI

$$XX' \rightarrow X^{\cdot} + X'^{\cdot}$$

XX'	X	X'	Effect of NO
CH_3COCH_3	CH_3	CH_3CO	Little[a]
$CH_3COC_2H_5$	CH_3	C_2H_5CO	Little[b]
$CH_3COC_3H_7$-n	CH_3	n-C_3H_7CO	No inhibition[c]
$C_2H_5COC_2H_5$	C_2H_5	C_2H_5CO	Inhibition[d]
CH_3CHO	CH_3	CHO	Little[e]
C_2H_5CHO	C_2H_5	CHO	Initial inhibition[f]

[a] Smith, J. R. E. and Hinshelwood, C. N. (1944). *Proc. Roy. Soc. A*, **183**, 33; [b] Waring, C. E. and Mutter, W. E. (1948). *J. Amer. Chem. Soc.* **70**, 4073; [c] Waring, C. E. and Garik, V. L. (1956). *J. Amer. Chem. Soc.* **78**, 5198; [d] Waring, C. E. and Barlow, C. S. (1949). *J. Amer. Chem. Soc.* **71**, 1519; [e] Freeman, G. R., Danby, C. J. and Hinshelwood, Sir C. N. (1958). *Proc. Roy. Soc. A*, **245**, 456; [f] Szabo, Z. G. and Marta, F. (1961). *J. Amer. Chem. Soc.* **83**, 768.

already noted that acyl radicals are likely to be σ and have surmised that methyl radicals can readily be formed in a similar configuration when the products of molecular decomposition; where these are the products of decomposition of the carbonyl compounds in Table VI, little inhibition results from the addition of NO.

On the other hand when ethyl, which as we have seen is more effectively stabilized in the π-form, is a product of decomposition the effect of adding NO is to cause a marked inhibition as a result of its scavenging the ethyl radical to form RNO· which is less reactive than the hydrocarbon radical. We have also noted that alkoxy radicals have a tendency to adopt the π-state and, true to the above predictions, the thermal decomposition of diethylether is strongly inhibited by the addition of NO[59]. But when the alkoxy radicals are produced in an excited, and therefore presumably σ-state by photolytic decomposition of peroxides and alkyl nitrites,[60] nitric oxide seems to have little effect on their decomposition.

As well as NO, olefins such as propene and hexene have been used as inhibitors of gas-phase-free radical pyrolysis reactions. We will see below that σ-radicals are efficient abstractors of hydrogen atoms so that it might be expected and indeed it is found that propene addition to the reactions of Table VI produces a certain amount of inhibition, the radical resulting from the abstraction from propene, namely allyl, being more stable than the products of the unhibited reaction.

Nitric oxide also acts as an inhibitor when added to the pyrolysis of a hydrocarbon but it does not halt the process completely, a maximally inhibited reaction remaining unaffected by the addition of further amounts of NO, although very large additions may induce a slight acceleration (Fig. 2).

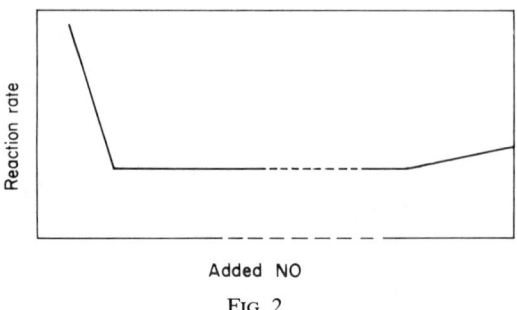

Isotopic labelling experiments have demonstrated[61] that the residual reaction is still a chain reaction involving free radicals and producing essentially the same products as the uninhibited process. Propene and other unsaturated hydrocarbons bearing a $-CH_2-$ group α to the unsaturated bond also inhibit such reactions to approximately the same extent although in general higher additions are required to produce the maximally inhibited reaction.

Nitric oxide is supposed to trap radicals,

$$R^\cdot + NO \rightarrow RNO^\cdot$$

and propene to inhibit via abstraction of a hydrogen atom from the methyl group to give the relatively stable allyl radical,

$$R^\cdot + CH_3CH=CH_2 \rightarrow RH + {}^\cdot CH_2CH=CH_2$$

and similarly for the other unsaturated inhibitors.

Many complex mechanisms have been suggested to explain the mechanism of inhibition[62] but a simple suggestion is that the inhibitors remove the radicals in the π-configuration leaving the σ-radicals to carry on the reaction. In the case of NO, when the energy absorbing property of the change in hybridization of the carbon atom bearing the unpaired electron and therefore forming a new bond with the nitrogen atom of the inhibitor, is sufficient to stabilize the resulting nitroso compound, the π-alkyl radical is removed. In the case of σ-radicals as we have seen there is no such stabilizing mechanism and in the absence of a highly efficient third body energy transfer mechanism the incipient nitroso compound redissociates leaving the alkyl radical free to attack the substrate molecule as in the uninhibited reaction and producing similar products.

The mechanism involving propylene and other similar olefins is not so obvious. As we have noted above a certain degree of inhibition is observed on the addition of propylene to carbonyl decompositions; these produce mainly σ-radicals so that this is one factor to be taken into account in the case of hydrocarbon pyrolysis. It is not, however, sufficiently efficient to explain the overall inhibiting effect of olefins and it may be suggested that the mechanism involves addition to the double bond of the π-radicals.[63] σ-Radicals as we shall see (page 112) also add efficiently to unsaturated bonds but, as with their addition to oxygen, if the resulting complex is not efficiently stabilized by energy transfer, it will readily redissociate.

To some extent this interpretation of the effect of inhibitors on hydrocarbon pyrolysis reactions rationalizes the inhibition mechanism put forward by Norrish and Pratt.[64] These authors identify μ-radicals as those involved in first order propagation steps and, as can be seen from the following mechanism proposed[65] for the pyrolysis of n-butane, they are larger than methyl and therefore more likely to be π.

$$C_4H_{10} \rightarrow 2C_2H_5^{\cdot}$$
$$\rightarrow C_3H_7^{\cdot} + CH_3^{\cdot}$$
$$C_4H_9^{\cdot} \rightarrow C_2H_4 + C_2H_5^{\cdot}$$
$$\rightarrow C_3H_6 + CH_3^{\cdot}$$
$$C_2H_5^{\cdot} + C_4H_{10} \rightarrow C_2H_6 + C_4H_9^{\cdot}$$
$$CH_3^{\cdot} + C_4H_{10} \rightarrow CH_4 + C_4H_9^{\cdot}$$
$$H^{\cdot} + C_4H_{10} \rightarrow H_2 + C_4H_9^{\cdot}$$
$$C_2H_5^{\cdot} \rightarrow C_2H_4 + H^{\cdot}$$
$$C_2H_5^{\cdot} + C_2H_5^{\cdot} \rightarrow n\text{-}C_4H_{10}$$
$$\rightarrow C_2H_4 + C_2H_6$$

It has also been noted by Gowenlock[62] that inhibition by propylene involving the addition of radicals to the double bond can be described by a mechanism similar to that of Norrish and Pratt.

The mechanism proposed here predicts that since the chains of the maximally inhibited reaction are carried by σ-radicals the A-factors of these residual reactions should be greater than those of the uninhibited reactions and, if initiation involves the dissociation of the hydrocarbon into two σ-radicals, the overall activation energies of the two processes should show the same trend. This has in fact been observed in the decompositions of alkanes,[66] certain ketones[67] and n-butanol.[68]

RADICAL–UNSATURATED BOND ADDITIONS

The main difficulty about a discussion of reactions of addition between organic free radicals and unsaturated molecules is the lack of data on A-factors and activation energies. Problems associated in measuring these Arrhenius factors arise from further reactions of the adduct radical and also from errors in the relatively low activation energies, errors which may be absolutely low but at the same time relatively large. Finally many addition reactions are exothermic giving rise to a radical with an excess of energy, a fact of little consequence if the subsequent energy transfer, either inter- or intra-molecular is efficient but of considerable importance if the excited radical either redissociates or reacts in an unexpected manner such as might occur if it behaved as a σ-radical. A discussion of some of the difficulties is given by Rabinovitch and Flowers.[69]

Not a great deal is known about the addition of σ-radicals to unsaturated systems; hydrogen atoms, however, have been found[70] to add with a low A-factor and small activation energy. Low temperature studies[71] have shown that the σ-radicals, phenyl, vinyl and CF_3, add to ethylene under conditions at which most alkyl radicals are unreactive towards olefins. An activation energy of $3350\,kcal\,mol^{-1}$ and a steric factor of 10^{-4} were calculated for the process,

$$C_2H_3^{\cdot} + C_2H_4 \rightarrow C_4H_7^{\cdot}$$

The low activation energy is presumably a reflection of the possibility of efficient overlap of the half-filled σ-orbital with the π-electron system of the unsaturated substrate producing a transition state with a structure close to that of the original reactants, *viz.*

(I)

Little or no rehybridization of the substrate or radical is involved in the formation of this complex and presumably unless the excess energy of the formation of the addition bond is dissipated for example by rapid rearrangement to final product, the complex redissociates. We have already noted this possibility as an explanation of the inhibiting effects of olefins on hydrocarbon pyrolysis. It is also clear that the complex will be fairly crowded; rotation and vibration will be hindered and, as found, the steric factor low.

A similar direct overlap mechanism in the case of a π-radical would produce considerable steric problems and it must be assumed that in the case of reactions of this sort the two reactants undergo a certain degree of rehybridization so that the transition state resembles the final product. more than the original species. The transition state is thus looser and the A-factor greater. Accompanying these changes will be the energy require- ment for rehybridization evident in an increase in the activation energy over the σ-radical addition process, as observed. Put in another way, the approach of a σ-radical to a double bond is sterically restricted to a direction perpendicular to the plane of the molecular framework (I) while overlap between a π-radical and a π-system both of which are distorting to σ-frame-

(II)

works has some affinity with the overlap of two σ-radicals and hence a larger A-factor (II). Once again the occurrence of a compromise to give the most favourable combination of Arrhenius factors vitiates attempts to calculate these parameters. In possible support of this latter type of transi- tion state, however, Bloor and his co-workers[72] have shown that the activa- tion energies for the addition of methyl and ethyl radicals to a series of olefins correlate with the localization energy—an indication that the transi- tion state resembles the products more than the reactants. On the other hand there has been considerable discussion[73] concerning the shape of radicals resulting from iodine atom addition to olefins and a planar transition state has been predicted for the addition to olefins of methyl and trifluoro- methyl, the latter of which is definitely a σ-radical. It is also interesting to note in this connection that in low temperature studies, no evidence was found[71] for the addition of $CCl_3^{\cdot}$ to ethylene under conditions where reaction with an activation energy less than about 5 kcal/mol^{-1} would have been detected. Once again we have a hint that the structure of $CCl_3^{\cdot}$ may be planar at low temperatures.

If the mechanism proposed above for radical addition to olefins is correct then we might expect that the A-factor would decrease with increased conjugation or hyperconjugation since both mechanisms will tend to oppose the π-bond rupture necessary to establish the looser transition state. By and large the results in Table VII support this conclusion. Conversely the A-factor ought to be larger if the resulting radical has a σ-configuration at the carbon atom bearing the unpaired electron, for

TABLE VII

Addition of radicals to unsaturated bonds[a]

Reaction	log A	E
$CH_3^{\cdot} + C_2H_4$	8·52	7·7
$+ C_3H_6$	8·22	7·4
$+ C_4H_8$-1	8·01	7·2
$+$ allene	8·3	8·0
$+$ butadiene	7·91	4·1
$+ C_2H_2$	8·4	7·7
$+ CH_3CCH$	8·7	8·8
$CF_3^{\cdot} + C_2H_4$	8·30	2·0
$+ C_3H_6$	8·02	0·54
$+ C_4H_8$-1(cis)	7·44	c. 0

[a] Kerr, J. A. and Parsonage, M. J. (1972). "Evaluated Kinetic Data in Gas Phase Addition Reactions", Butterworths, London.

example in reactions involving $CF_3^{\cdot}$ addition to acetylenes giving vinyl radicals[74] or the reaction,

$$R^{\cdot} + CH_2{=}CF_2 \rightarrow RCH_2{-}CF_2^{\cdot}$$

and this is very largely borne out by recent work of Tedder and colleagues who obtained the following figures for the addition of methyl to a series of olefins,[75]

CH$_2$		CHF		CF$_2$	
log A	E	log A	E	log A	E
$CH_2{=}CH_2$ 8·6	3·4	$CHF{=}CH_2$ 9·5	9·9	$CF_2{=}CH_2$ 7·9	8·2
$CH_2{=}CFH$ 9·2	8·8	$CHF{=}CF_2$ 8·2	8·6	$CF_2{=}CFH$ 7·8	6·2
$CH_2{=}CF_2$ 9·8	11·2			$CF_2{=}CF_2$ 7·7	5·3
$CH_2{=}CHCl$ 8·4	9·8				
$CH_2{=}CHCCl_3$ 7·9	7·1				

and similarly for the addition of trichloromethyl:[76]

CH$_2$		CHF		CF$_2$	
log A	E	log A	E	log A	E
$CH_2{=}CH_2$ 5·6	3·2	$CFH{=}CH_2$ 5·4	5·4	$CF_2{=}CH_2$ 5·5	8·3
$CH_2{=}CHF$ 5·5	3·3	$CFH{=}CF_2$ 6·3	6·1	$CF_2{=}CHF$ 6·4	7·1
$CH_2{=}CF_2$ 5·7	4·6			$CF_2{=}CF_2$ 7·1	6·1

The part structures at the head of each column indicate the olefin end at which addition is occurring. Both sets of figures are complicated by the effects noted by Kilcoyne and Jennings and noted at the end of Chapter 1 and also in the latter case by the electrophilicity of the CCl_3 radical, but it can be seen that in general the greater the tendency of the resulting adduct radical to be σ, i.e. the greater the degree of fluorine substitution of the carbon atom bearing the unpaired electron, the greater the A-factor.

An interesting example of radical addition is the reaction between allene and hydrogen bromide[76] which proceeds via bromine atom addition as a first step:

$$Br^{\cdot} + CH_2{=}C{=}CH_2 \rightarrow CH_2{=}CBr{-}CH_2{}^{\cdot} \xrightarrow{\text{HBr}} CH_2{=}CBr{-}CH_3$$

or

$$\rightarrow CH_2{=}\overset{\cdot}{C}{-}CH_2Br \xrightarrow{\text{HBr}} CH_2{=}CH{-}CH_2Br$$

Terminal addition give the vinyl radical (III) while central addition of the bromine atom gives the allyl radical (IV),

$$(III) \qquad\qquad (IV)$$

The formation of the latter which is resonance stabilized is likely to be efficient in dissipating the excess energy of the new C—Br bond whereas the structure of the former, being more like the original molecule is less able to accommodate this extra energy and more likely to redissociate in the absence of highly efficient energy transfer by a third body or of a substrate with a readily abstractable hydrogen atom. True to this conclusion, reaction in the presence of low concentrations of HBr proceeds predominantly by central addition. Increase in the concentration establishes the 2:1 terminal to central addition ratio expected on statistical grounds and exceeds it probably because hydrogen abstraction by the vinyl radical from the HBr is more efficient than that by the allyl radical.

On the basis of the above mechanism we can also explain the stereo-specificity of the addition of HBr to olefins, a stereospecificity which tends to disappear as the temperature of the reaction medium is increased. Since bromine atoms are formally σ-radicals, they will add efficiently in a direction perpendicular to the plane of the double bond,

At low temperatures the adduct radical appears to be sufficiently long-lived to abstract a deuterium atom from DBr, the direction of addition of the D· atom being for steric reasons opposite to that of the bromine, as follows, giving a product with a preferred geometry,

As the temperature is raised, the initially formed planar complex will be able the more quickly to achieve its stable π-type structure with relatively unhindered rotation and there will be no preferred direction of addition. DBr and HBr are particularly good transfer agents, a factor which increases the chance of the planar complex abstracting before rearrangement to the stable state.

AROMATIC SUBSTITUTION

Aromatic substitution is one of the more commonly studied of chemical reactions and as we noted in Chapter 1 it is now established that homolytic substitution follows broadly the mechanism,

although the details of reaction and particularly of the effects of substituents on the nucleus of the substrate are not well understood. The overall picture, as illustrated in Table VIII is quite different from that observed in the cases of electrophilic or nucleophilic substitution. On a purely statistical basis we would anticipate finding o-, m- and p-substituted products of radical attack in the quantitative ratio 2:2:1 respectively but, although the ratios of m:p substituted products are often as expected, reaction tends to take place predominantly at the o- position.

Several possible reasons can be put forward to attempt to explain the trends of the results in Table VIII.[77]

TABLE VIII

Addition of radicals to aromatic compounds[a]

| | Percentage Products | | |
	o	m	p
$C_6H_5^{\cdot} + C_6H_5CH_3$	66·5	19·3	14·2
$+ C_6H_5C_2H_5$	53·0	28·0	19·0
$+ C_6H_5C_3H_7\text{-i}$	31·0	42·0	27·0
$+ C_6H_5C_4H_9\text{-t}$	24·0	49·0	27·0
$+ C_6H_5C_6H_5$	48·5	23·0	28·5
$CH_3^{\cdot} + C_6H_5CH_3$	56·5	26·5	17·0
$+ C_6H_5Cl$	64·0	25·0	11·0
$+ C_6H_5Br$	67·5	23·0	9·5
$+ C_6H_5OCH_3$	74·0	15·0	11·0

[a] Williams, G. H. (1964). "Homolytic Aromatic Substitution", Pergamon Press, London.

(1) Mesomeric Effects

Such effects seem to be important in substitution in benzonitrile and possibly anisole where the o-, p-directing effect can be explained by resonance stabilization of structures of the sort,

2. Steric Effects

There are two distinct steric effects. The first is a consequence of the size of the substituent on the benzene ring and/or of the attacking radical. As Table VIII shows when either or both of these is large there is a reduced yield of o-substituted product, an extreme case being that of cyclohexyl substitution of t-butyl benzene.

The second effect derives from the configuration of the substituted cyclohexadienyl radical. Small cycloalkyl radicals tend to relieve ring strain by adopting a σ-configuration. Cyclohexadienyl is probably also strained and it is not unlikely that there is a tendency to twist the structure from the planar and to introduce a contribution from the carbon 2s orbital into the p orbital containing the unpaired electron.[78]

However, as we have seen (Chapter 3, Table VI), the cyclohexadienyl radical has a resonance energy close to 20 kcal mol^{-1} and it is likely that the unsubstituted radical will be planar, a conclusion which is supported by e.s.r. results.[79] Now suppose a substituent is introduced into this planar ring. Further strain is introduced by interference between the substituent and the *o*-hydrogen atom,

Relief of this strain would be possible were there some mechanism capable of competing with the resonance energy and giving some extra stability to the configuration in which the carbon atom bearing the unpaired electron adopts a non-planar configuration. In Chapter 2 we have discussed the four factors which contribute to the energetics of a free radical. Of these we can ignore in the present context the valence state excitation energy and bond energy considerations. Resonance energy we have already mentioned. Electronegative effects could play an important part in stabilizing non-planar structures under favourable conditions; for example with a highly electronegative substituent we might expect contributions from structures of the sort,

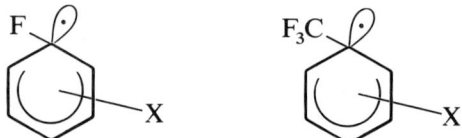

But it is likely that the structure which has the least strain is that which permits both substituents, the original and that after the initial radical substitution, to lie out of the plane of the ring. If such a configuration is written for the *o*-substituted isomer then it is clear immediately that extra stability can be achieved by the formation of a 3-centre bond by the overlap of the σ-orbital containing the unpaired electron and the orbital of the hydrogen atom attached to the carbon atom which has been attacked, thus

Such a combination of relieved ring and steric strain and reduced free electron energy is possible only in the case of the *o*-substituted isomer and is suggested to be the reason for the preponderance of *o*-substituted benzenes in the products of such radical substitution reactions.

HYDROGEN ABSTRACTION REACTIONS

Hydrogen abstraction reactions involve the approach of a free radical and a hydrogen containing substrate and subsequent reaction to produce a new radical and a new molecule,

$$R^{\bullet} + R'H \rightarrow RH + R''$$

We will ignore in this section hydrogen abstraction from a second radical since this is (probably) disproportionation and has already been discussed. A full discussion of hydrogen abstraction reactions, again ignoring polar effects, requires a knowledge of the configurations of both $R^{\bullet}$ and R'' but once again we shall simplify the situation by taking $R^{\bullet} = CH_3^{\bullet}$ and considering only the effects of varying R'. Arrhenius parameters for reactions of this sort are usually measured in a system where there is competition between

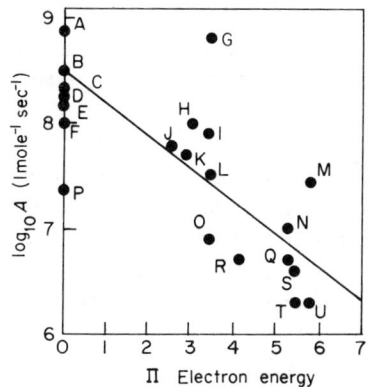

Fig. 3. Metathetical reactions of methyl radicals; plot of $\log_{10} A$ against the π-electron energy of the radicals produced, A, CH_3CO;[a] B, H;[b] C, cyclo-C_4H_7;[c] D, cyclo-C_3H_5;[d] E, CH_3;[e] F, CHO;[f] G, CH_2OOCH_3;[g] H, CH_2CN;[h] I, CH_2OH;[i] J, $CH_2NHNHCH_3$;[j] K, $CH_2CH=CH_2$;[k] L, CH_2NH;[l] M, CH_3O;[m] N, CH_3CH_2;[m] O, CH_2OH;[m] P, CF_3;[n] Q, CH_3NNH_2;[i] R, CH_2-CHNNCHCH$_3$;[l] S, CH_3NNHCH_3;[j] T, CH_3NH;[j] U, CH_3O.[i]

[a] Brinton, R. K. and Volman, D. H. (1954). *J. Chem. Phys.* **22**, 929. [b] Whittle, E. and Steacie, E. W. R. (1953). *J. Chem. Phys.* **21**, 993; Phibbs, M. K. and Darwent, B. de B. (1949). *Trans. Faraday Soc.* **45**, 541. [c] Gordon, A. S., Smith, S. R. and Drew, C. M. (1962). *J. Chem. Phys.* **36**, 824. [d] Gordon, A. S. and Smith, S. R. (1962). *J. Phys. Chem.* **66**, 521. [e] Dainton, F. S., Ivin, K. J. and Wilkinson, F. (1959). *Trans. Faraday Soc.* **55**, 929. [f] Blake, A. R. and Kutschke, K. O. (1959). *Canad. J. Chem.* **37**, 1462; Toby, S. and Kutschke, K. O. (1959). *Canad. J. Chem.* **37**, 672. [g] Thynne, J. C. J. and Gray, P. (1963). *Trans. Faraday Soc.* **59**, 1149.[h] Wijnen, M. H. J (1954). *J. Chem. Phys.* **22**, 1074. [i] Shannon, T. W. and Harrison, A. S. (1963). *Canad. J. Chem.* **41**, 2455. [j] Gray, P., Herod, A. A., Jones, A. and Thynne, J. C. J. (1966). *Trans. Faraday Soc.* **62**, 2774. [k] Miyoshi, M. and Brinton, R. K. (1962). *J. Chem. Phys.* **36**, 3019. [l] Gray, P. and Thynne, J. C. J. (1963). *Trans. Faraday Soc.* **59**, 2275; Brinton, R. K. (1960). *Canad. J. Chem.* **38**, 1339. [m] Shaw, R. and Thynne, J. C. J. (1966). *Trans. Faraday Soc.* **62**, 104. [n] Pritchard, G. O. and Thommarson, R. L. (1964). *J. Phys. Chem.* **68**, 568.

abstraction and combination, *viz.*

$$CH_3^{\cdot} + RH \rightarrow CH_4 + R^{\cdot} \tag{6}$$

$$2CH_3^{\cdot} \rightarrow C_2H_6 \tag{7}$$

A measure of the concentrations of methane and ethane gives the ratio of the rate constants $k_6/k_7^{\frac{1}{2}}$ and if k_7 is known k_6 is readily determined. k_7 for radicals as we have seen is a quantity whose value may not be well known but if different workers use the same value then it is possible to derive a series of values of k_6 which, though not absolutely accurate are comparable.

Again, as can be seen from Fig. 3, the suggestion that A-factors decrease with increasing π-stability of the product radical, is justified, i.e. the complex involved in the abstraction process is tighter the more the inchoate radical is stable in the π-configuration. Figure 3 is a plot of $\log_{10} A$ for the reaction,[1]

$$CH_3^{\cdot} + RH \rightarrow CH_4 + R^{\cdot}$$

against the π-electron energy of $R^{\cdot}$; since most of the radicals produced have only two carbon atoms the concept of the delocalization energy is not appropriate. The full line in the figure has the equation,

$$\log_{10} A = (8\cdot5 - 0\cdot32) \pm 0\cdot5$$

where A is in units of $l.\,mol^{-1}\,s^{-1}$ and e is the π-electron energy of $R^{\cdot}$ in units of β. This equation does not include the results for the radicals produced by abstraction from dimethylperoxide and fluoroform. Again as we have seen in our discussion of dissociations, the production of amino radicals is accompanied by low A-factors. This has also been shown recently in work on ethyleneimine,[80]

	log A	E (kcal mol^{-1})
$^{\cdot}CH_3 + (CH_2)_2NH \rightarrow CH_4 + (CH_2)_2N^{\cdot}$	7·17	4·6
$\rightarrow CH_4 + \underset{\diagdown\ \ \diagup}{\underset{NH}{CH_2{-}CH^{\cdot}}}$	8·44	10·1

By analogy with cyclopropyl the radical produced in the latter reaction might well be a σ-radical and the higher A-factor is explained. Similar changes in A-factors have been obtained in the following cases,[81]

	log A	E (kcal mol^{-1})
$CH_3^{\cdot} + N_2H_4 \rightarrow CH_4 + N_2H_3^{\cdot}$	11·0	5·6
$CH_3^{\cdot} + CH_3NHNHCH_3 \rightarrow CH_4 + CH_3\dot{N}NHCH_3$	9·9	2·1
$CH_3^{\cdot} + CH_3NDNDCH_3 \rightarrow CH_4 + CH_3NDNDCH_2^{\cdot}$	11·6	6·6

Morris and Thynne[82] also found a similar effect in the abstraction reactions of $CF_3^{\cdot}$ with methylamine.

In the case of alcohols it has now been established[83] that abstraction from the hydroxy group to give the π-alkoxy radical has a low A-factor whereas that from the α-CH_2R group (R = H or alkyl) is fairly high,[84] in fact higher than would be expected particularly if hyperconjugation were a property of the radical produced. This could be explained if the free electron in the radicals produced in the abstraction has some σ-character, a possibility which as we have seen above does not conflict with e.s.r. data on these radicals.[85]

The high A-factors for abstraction reactions involving the carbonyl group of acetaldehyde are also explicable on the basis of a σ-structure for the carbonyl radical[86] (see page 55).

These effects are not restricted to aliphatic compounds; abstraction of hydrogen atoms from m-substituted toluenes is accompanied by a higher A-factor than with the corresponding o- and p-substituted compounds.[87] The unpaired electrons in the radical products from the latter compounds have a greater degree of delocalization, canonical forms of the sort,

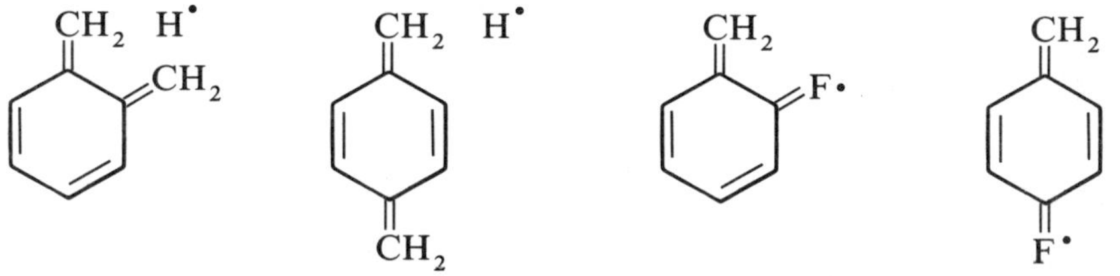

being impossible in the case of the m-substituted substrates.

At first sight it might be hoped that detailed studies of isotope effects in hydrogen abstraction reactions would be able to shed some light on the mechanism of hydrogen abstraction. Take, for example, the system,

$$R^{\cdot} + SH \rightarrow (R \cdots H \cdots S) \rightarrow RH + S^{\cdot}$$

$$R^{\cdot} + SD \rightarrow (R \cdots D \cdots S) \rightarrow RD + S^{\cdot}$$

Because of zero point differences the S—D bond is some $2\,kcal/mol^{-1}$ stronger than the C—H bond so that if the transition state were to involve the breakage of the S—H and S—D bonds, the former reaction would require the smaller amount of energy and there would be a primary isotope effect measured as k_H/k_D which is found experimentally to vary from unity to about eight, the larger values apparently reflecting a greater stretching of the S—H bond in the transition state. The true position, however, looks to be somewhat more complex than this simple explanation and it has

E

been suggested[88] that the high values obtained for this ratio can be explained on the basis of a tunnel effect. This is illustrated by reference to Fig. 4. Hydrogen atoms, being lighter than deuterium atoms are more easily able to tunnel through rather than climb over the activation barrier.

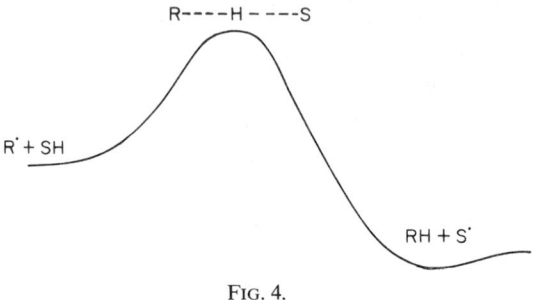

FIG. 4.

We have noted above that increasing the temperature of a radical system is one way in which to encourage the formation of σ-radicals and it might be reasonable to suppose that the activation energy of an abstraction reaction should increase with increasing temperature. Table IX shows that such an effect is observable in the case of abstraction by a hydrogen atom,[89] and it has recently also been found[110] for abstraction reactions of methyl and ethyl radicals.

TABLE IX

Variation of Arrhenius parameters with temperature
$$H^{\cdot} + CH_4 \rightarrow H_2 + CH_3^{\cdot}$$

$\log A\,[(\text{l. mol}^{-1}\,\text{s}^{-1})]$	E_A (kcal mol^{-1})	T°K
7·0	4·5	372–436
9·5	9·0	403–503
10·93	14·1	773–873
10·08	10·6	843–933
11·3	11·5	1100–1900

It is also of interest in this context that information about activation energies of metathetical reactions can be derived using simple Hückel theory on the substrate, but in this case on the σ-framework.[90] Figure 5 is a plot of the activation energy for methyl radical abstraction reactions against the deviation from unity of the electron density at the hydrogen atom being attacked. The activation energy, measured in kcal mol^{-1} is given by the relationship,

$$E_A = [16 - 70(1 - 2c_1^2)] \pm 2$$

where c_1 is the atomic orbital coefficient for the hydrogen atom in question. Propene lies outside these limits presumably because the present treatment takes no account of hyperconjugation but the small divergence is in line with the opinion[91] that hyperconjugation effects are small compared with changes in the hybridization where bond dissociation energies of non-aromatic hydrocarbons are concerned. This is of course the basis on which the bulk of the calculations in Chapter 3 were performed.

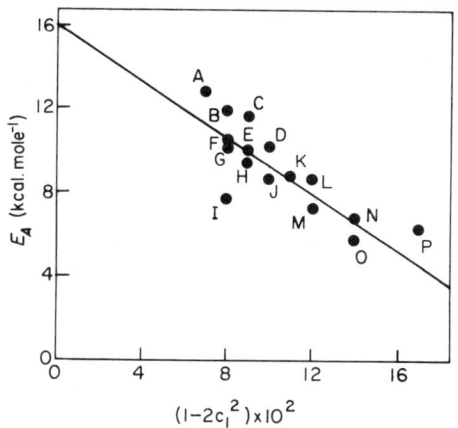

$$(1-2c_1^2) \times 10^2$$

Fig. 5. Metathetical reactions of methyl radicals; plot of the activation energy against the electron density on the hydrogen atom being removed. A, CH_4; [a] B, C_2H_6; [b] C, C_4H_{10}; [c] D, CH_3Br; [d] E, C_2H_4; [e] C_4H_{10}; [c,f] F, C_2H_6; [e] G, C_3H_8; [g] H, CH_3Cl; [d] I, C_3H_6; [e] J, C_4H_6; [h] K, CH_3F; [d] L, CH_2Br_2; [d] M, CH_2Cl_2; [d] N, $CHCl_3$; [i] O, $CHCl_3$; [d] P, CH_2F_2.[d]

[a] Shaw, R. and Thynne, J. C. J. (1966). *Trans. Faraday Soc.* **62**, 104; [b] McNesby, J. R. and Gordon, A. S. (1955). *J. Amer. Chem. Soc.* **77**, 4719; [c] McNesby, J. R. and Gordon, A. S. (1956). *J. Amer. Chem. Soc.* **78**, 3570; [d] Raal, F. A. and Steacie, E. W. R. (1952). *J. Chem. Phys.* **20**, 578; [e] Trotman-Dickenson, A. F. and Steacie, E. W. R. (1951). *J. Chem. Phys.* **19**, 169; [f] Tedder, J. M. and Watson, R. A. (1966). *J. Chem. Soc. B*, 1069. [g] Jackson, W. M., McNesby, J. R. and Darwent, B. de B. (1962). *J. Chem. Phys.* **37**, 1610; [h] Trotman-Dickenson, A. F. and Steacie, E. W. R. (1951). *J. Chem. Phys.* **19**, 329; [i] Cvetanovic, R. J., Raal, F. A. and Steacie, E. W. R. (1953). *Canad. J. Chem.* **31**, 171.

Turning now to reactions involving σ-radicals, because of the highly directional qualities of σ-orbitals these radicals should readily abstract hydrogen atoms from organic substrates. It has been found for example, that phenyl, vinyl, cyclopropyl and $CF_3^{\bullet}$ will abstract hydrogen atoms from saturated hydrocarbons at 77°K,[92] conditions under which methyl and other π-radicals are unreactive. Other authors have shown that phenyl abstracts more readily than methyl but less than $CF_3^{\bullet}$.[93] As we have already noted, succinimidyl tends to abstract rather than to combine and the hydroxyvinyl radical, formed in the oxidation of acetylene,[50] will abstract even the tightly held acetylene proton,

$$HC\equiv CH + {}^{\cdot}OH \rightarrow H\overset{\cdot}{C}=CH(OH) \xrightarrow{C_2H_2} H_2C=CH(OH)$$

in competition with the (inefficient) addition reaction with oxygen.

Of more interest are the various possible intramolecular abstraction reactions which have been observed with vinyl radicals. We shall consider these in order of the size of the ring in the transition state.

1,5- HYDROGEN ABSTRACTION

Heiba and Dessau have shown[94] such reactions to occur in the reactions following the addition of $CCl_3{}^{\cdot}$ radicals to various acetylenes, the overall reaction scheme being,

The activation energy for the 1,5 shift (E_H) (reaction 10) is slightly greater than that for chlorine abstraction (reaction 9),

$$E_H - E_{Cl} = 2\cdot7 \pm 0\cdot5 \text{ kcal mol}^{-1}$$

but comparison of the relative ratio of 1,5-hydrogen shift to chlorine transfer in the case of the secondary vinyl radical with the corresponding ratio reported for the primary alkyl radical $Cl_3C(CH_2)_5CH_2\cdot$, after correction to the same temperature, leads to the value for,

$$(k_H/k_{Cl})_{vinyl}/(k_H/k_{Cl})_{primary} = c.\ 560$$

thus demonstrating the very high reactivity in abstraction of the vinyl radical which is probably σ. The selectivity of abstraction is also high being 1:22:650 for the abstraction of primary, secondary and tertiary hydrogen atoms respectively. This is explained by the fact that the radical moiety cannot approach the hydrogen atom to be abstracted along the line of the C—H bond because of the cyclic nature of the transition state; this increases the activation energy, an effect which has also been shown for analogous reactions of alkoxy radicals.[95]

The cyclization reaction (11) is faster than abstraction of chlorine from carbon tetrachloride and of hydrogen from thiols by the alkyl radical. A further measure of the speed of cyclization is the retention of a part of the optical activity of the ester (V) in the lactone (VI) formed on cyclization,

$$[\alpha]_D^{25} = +20\cdot6°$$
(V)

$$[\alpha]_D^{25} = +0\cdot9°$$
(VI)

Normally when an optically active site at a carbon atom becomes a radical site the activity is lost due to the flattening of the bonds around the trigonal carbon atom and the formation of a π-radical. In the reaction considered here it looks as though the active site has retained much of its original configuration around the carbon atom on the loss of the hydrogen atom. It is suggested that the reason for this lies in the nature of the hydrogen abstraction process by the σ-radical analogous to reaction (10) above. Such an abstraction, like the dissociation of an organic compound into two σ-radicals, is likely to involve no substantial molecular rearrangement on the formation of the transition state; put in another way, the directional quality of the σ-orbital containing the unpaired electron in the σ-radical is such that abstraction occurs at a greater distance than is the case with a π-radical. It follows therefore that if the alkyl radical so formed can cyclize before rearrangement to the stable π-state, the lactone will retain a degree of optical activity.

1,4-HYDROGEN ABSTRACTION

1,4-abstraction has been observed by Trotman-Dickenson[96] following the addition of isopropyl and t-butyl radicals, produced by the photolysis of the corresponding aldehydes, to acetylene and propyne. The products from iso-propyl and acetylene include pentene-1, the formation of which can be explained by the sequence,

$$CH_3CH_2\dot{C}HCH=CH_2 \xrightarrow{RH} CH_3CH_2CH_2CH=CH_2$$

The vinyl radical when initially formed will have excess of energy, a factor which no doubt helps to drive the 1,4-abstraction reaction; indeed when inert gases which can absorb this energy are added the isomerization is inhibited.

The corresponding reaction with ethyl cannot be demonstrated in such a simple system since the product is the same as that which would be formed via an intermolecular hydrogen abstraction,

$$CH_3CH_2CH=CH_2$$

Cyclization of the alkyl radical to the corresponding cyclobutane has not been observed but a similar sequence has been proposed[50] for the formation of formaldehyde in acetylene oxidation,

1,3-HYDROGEN ABSTRACTION

1,3-abstraction has not been observed following the addition of methyl radicals to acetylene or propyne but it has been postulated[50] in the mechanism for the gas phase oxidation of acetylene to account for the formation of ketene as a primary product,

The four-membered ring transition state is highly strained and its formation not very probable. In most systems alternative and more favourable reactions would swamp the possible further reactions of such a species but in the oxidation under consideration there are few if any alternative reaction pathways which are energetically more likely and it is possible to observe this rather unusual process. The cyclization reaction to give ethylene oxide has not been recorded.

"HOT" RADICALS

This chapter ends with a very brief discussion of the reactions of "hot" radicals produced by high energy techniques and specifically by the photolysis of alkyl iodides,

$$R\!-\!I \xrightarrow{\ h\nu\ } R^{\cdot *} + I^{\cdot *}$$

Where $\nu = 2537\,\text{Å}$ and $R^{\cdot}$ is methyl the results indicate that the iodine atom is formed mainly in the excited $^2P_{\frac{1}{2}}$ state and the methyl radical with some $32\,\text{kcal mol}^{-1}$ excess energy which increases to $35\text{--}45\,\text{kcal mol}^{-1}$ when the wavelength of the incident light is shortened to $2288\,\text{Å}$.

The reactions of such "hot" methyl radicals can be studied in the presence of a large number of types of organic compounds, hydrocarbons, alcohols, esters etc, all of which are transparent at these wavelengths. As discussed by Bass and Pimentel,[97] the electronic excitation of methyl iodide at these wavelengths is such that the out-of-plane bending modes of the resulting methyl radical, in accordance with the Frank–Condon principle are excited. These authors estimate the excitation energy at about $10\,\text{kcal mol}^{-1}$ but the rather lower value of $5\,\text{kcal mol}^{-1}$ deduced in Chapter 3 is closer to the excitation energy observed recently in work on MeI photodissociation.[111] In other words "hot" methyl produced in this way is a σ-radical and in the systems indicated above we should expect its reactions to confirm the predictions made as to the likely reactions of such a radical elsewhere in this book.

The particular reaction systems and experimental methods are not, unfortunately, amenable to the acquisition of Arrhenius data nor, as yet, can the products of the reactions be ascribed to hot-radical–hot-radical reactions but an increasing amount of information concerning the hydrogen abstraction reactions of hot radicals is becoming available in such a form that it can be compared to that from analogous reactions involving cold, or thermally equilibrated, methyl radicals. Under normal reaction conditions and in the absence of highly polar hydrogen donors the hydrogen abstraction reaction kinetics of methyl and trifluoromethyl are roughly similar no doubt due to the close similarity of the values of $D(CH_3\!-\!H)$ and $D(CF_3\!-\!H)$. Hot methyl, on the other hand, is much more reactive than trifluoromethyl.[98] Both radicals produced by photodissociation contain excess energy distributed amongst their translational and vibrational modes but they differ in one essential respect—the ground state of trifluoromethyl is pyramidal while that of methyl is planar so that should the deformation vibrations of the two "hot" radicals be excited, trifluoromethyl will present itself to a substrate containing an abstractable hydrogen atom in a configuration similar to that of its ground state albeit with excess vibrational energy while methyl will look to the substrate as a σ-radical and the strength of the bond resulting from abstraction will be greater by at least 5 kcal mol^{-1} (see Chapter 3) than that formed with CF_3˙ radical. The hot methyl radical will therefore be more reactive. Of course, in view of its greater inertia, hot trifluoromethyl may also have a smaller velocity than hot methyl which may also have something to do with its smaller activity.

A number of other predictions also arises from the suggestion that hot methyl radicals are σ. For example, they should be inefficiently scavenged by oxygen and nitric oxide (page 103) and indeed this has been shown to be the case.[99, 100] Further, in hydrogen abstraction reactions, if the resulting radical remains in the tetrahedral configuration as we discussed above in the case of intra-molecular hydrogen abstraction by vinyl radicals, there should be neither inter- not intra-molecular isotope effects and this has been noted in the solid state.[100] Also under such conditions the selectivity of the abstraction process should not be as high as in those cases where the resulting radical can assume the stable π-configuration and again this has been demonstrated.[100] The ratio rate$_{primary}$:rate$_{secondary}$:rate$_{tertiary}$ for hydrogen atom abstraction in the solid state is about $1:2\cdot2:2$ compared to the corresponding ratio of $1:3\cdot6:10$ found in the gas phase.[101]

Hot CD_3˙ is less reactive than hot CH_3˙, C_2D_5˙ than C_2H_5˙ and C_2H_5˙ than CH_3˙ and it has been suggested[102] that this is due to the more efficient absorption of excess energy in modes other than translational, the distribution of which is supposed to be the cause of the enhanced reactivity of the hot radicals. Ethyl has more degrees of freedom than methyl and the spacing of the vibrational energy levels of the deuteriated radicals being closer together the excess energy can be more diffusely accommodated. Such a suggestion is inherently reasonable but it can also be seen that by increasing the mass of groups attached to the trigonal carbon atom the height of the inversion barrier (Chapter 2, Fig. 4) is increased and more of the radicals in the excited state will be trapped in the central well and, in addition to being inherently less reactive than the σ-state of the radical, are liable to scavenging by, for example, oxygen.

Simons and his colleagues[103] have shown that hot halogenated methyl radicals are better hydrogen abstractors than the corresponding thermalized radicals and suggest vibrational excitation to be responsible since addition of argon, with a mass close to that of for example CH_2Cl˙ does not affect the quantum yield of the formation of methyl chloride in the reaction sequence,

$$CH_2ClBr \xrightarrow{h\nu} \ ^{\cdot}CH_2Cl + Br^{\cdot}$$

$$\xrightarrow{CH_2ClBr} CH_3Cl + \ ^{\cdot}CHClBr.$$

Were the excitation energy facilitating the hydrogen abstraction process translational, then we would expect argon rapidly to thermalize the radicals and reduce the quantum yield of the formation of methyl chloride.

Added inert gases have been shown to quench hot $C_2D_5^{\cdot}$,[102] the relative efficiencies being He—0·50, Ne—0·62, N_2—1·00 and CO_2—1·51, the same sort of order found to be effective in quenching explosions in the slow combustion of acetylene, explosions which were induced by the addition of inert gas in the first place.[104] The more complex the molecule of added gas the more efficient the energy absorption. The oxidation of acetylene is supposed[50] to proceed via σ-radicals which add oxygen in a highly reversible reaction,

the vibrationally excited radical being quenched by acetylene,

Added inert gas replaces acetylene in the second step and if it is less efficient than the acetylene it is probable that there will be a build up of highly excited vinyl radicals leading to explosion.

Simons suggested[105] that the increase in the reactivity in hydrogen abstraction reactions of hot radicals implies the formation of a long-lived collision complex between the radical and the thermalized substrate during the lifetime of which efficient energy transfer can occur. Strongly coupled collision complexes are envisaged by Magee and Hamill[106] in their absolute rate theory of hot radical reactions and by Light[107] in his statistical theory but no such artificial assumption is necessary in the theory presented here of hot-radicals acting as σ-radicals.

REFERENCES

1. Hay J. M. (1967). *J. Chem. Soc. B*, 1175.
2. Streitwieser, A. Jr. (1961). "Molecular Orbital Theory for Organic Chemists", Wiley, New York.
3. Molyneux, P. (1966). *Tetrahedron* 22, 2929.
4. Kerr, J. A., Sekhar, R. C. and Trotman-Dickenson, A. F. (1963). *J. Chem. Soc.* 3217.
5. Yee Quee, M. J. and Thynne, J. C. J. (1968). *J. Phys. Chem.* 72, 2824.
6. Cox, D. L., Livermore, R. A. and Phillips, L. (1966). *J. Chem. Soc. B*, 245.
7. Gray, P., Shaw, R. and Thynne, J. C. J. (1967). *Progr. Reaction Kinetics* 4, 63.
8. Legett, C. and Thynne, J. C. J. (1970). *J. Chem. Soc. A*, 1188.
9. Rust, F. R., Seubold, F. M. and Vaughan, W. F. (1950). *J. Am. Chem. Soc.* 72, 338.
10. Hershenson, H. and Benson, S. W. (1962). *J. Chem. Phys.* 37, 1889.
11. Lin, M. C. and Laidler, K. J. (1967). *Canad. J. Chem.* 45, 1315.
12. Lin, M. C. and Laidler, K. J. (1966). *Canad. J. Chem.* 44, 2927.
13. Lin, M. C. and Back, M. H. (1966). *Canad. J. Chem.* 44, 2369.
14. O'Neal, H. E. and Benson, S. W. (1962). *J. Chem. Phys.* 36, 2196.
15. Rabinowitch, B. S., Kubin, R. F. and Harrington, R. E. (1963). *J. Chem. Phys.* 38, 405.
16. Grotewold, J., Lissi, E. A. and Neumann, M. G. (1968). *J. Chem. Soc. A*, 375.
17. Casas, F. Previtali, C., Grotewold, J. and Lissi, E. A. (1970). *J. Chem. Soc. A*, 1001.
18. Rabinowitch, B. S. and Setzer, D. W. (1964). *Adv. Photochem.* 3, 1.
19. Krech, M. and Price, S. J. W. (1967). *Canad. J. Chem.* 45, 157; Loucks, L. F. (1967). *Canad. J. Chem.* 45, 2775; Shaw, H., Menczel, J. H. and Toby, S. (1967). *J. Phys. Chem.* 71, 4180; Liu, M. T. H. and Laidler, K. J. (1968). *Canad. J. Chem.* 46, 479; Hole, K. J. and Mulcahy, M. F. R. (1968). *J. Phys. Chem.* 73, 177; Shaw, H. and Toby, S. (1968). *J. Phys. Chem.* 72, 2337.
20. Frey, H. M. and Walsh, R. (1969). *Chem. Rev.* 69, 103.
21. Waage, E. V. and Rabinowitch, B. S. (1971). *Intern. J. Chem. Kinetics* 3, 105.
22. Hase, W. L. and Simons, J. W. (1971). *J. Chem. Phys.* 54, 1277.
23. O'Neal, H. E. and Benson, S. W. (1969). *Intern. J. Chem. Kinetics* 1, 221.
24. Pacey, P. D. and Purnell, J. H. (1972). *J. Chem. Kinetics* 4, 657.
25. Terry, J. O. and Futrell, J. M. (1968). *Canad. J. Chem.* 46, 664.
26. Kerr, J. A. and Trotman-Dickenson, A. F. (1959). *J. Chem. Soc.* 1602.
27. Tedder, J. M. and Walton, J. C. (1966). *Chem. Comm.* 140.
28. Melville, H. W., Robb, J. C. and Tutton, R. C. (1951). *Disc. Faraday Soc.* 10, 154; (1953). 14, 150; Bengough, W. S. and Thomson, R. A. M. (1961). *Trans. Faraday Soc.* 57, 1928.
29. Eckling, R., Goldfinger, P. Huybrechts, G., Martens, G., Meyers, L. and Smoes, S. (1960). *Chem. Ber.* 93, 3104.
30. Beck, P. W., Kniebes, D. V. and Gunning, H. E. (1954). *J. Chem. Phys.* 22, 678.
31. McNesby, J. R., Drew, C. M. and Gordon, A. S. (1960). *J. Phys. Chem.* 59, 988.
32. Cadman, P., Ivel, Y. and Trotman-Dickenson, A. F. (1970). *J. Chem. Soc. A*, 1207.
33. *See* Hoare, D. E. and Pearson, G. S. (1964). *Adv. Photochem.* 8, 83.
34. Benson, S. W. (1965). *J. Amer. Chem. Soc.* 87, 972.
35. McKellar, J. F. and Norrish, R. G. W. (1961). *Proc. Roy. Soc. A*, 263, 51.

36. Barber, M., Farren, J. and Linnett, J. W. (1963). *Proc. Roy. Soc.* A, **274**, 306.
37. Bradley, J. N. (1961). *J. Chem. Phys.* **35**, 748; Bradley, J. N. and Rabinowitch, B. S. (1962). *J. Chem. Phys.* **36**, 3498.
38. Kominar, R. J., Jacko, M. G. and Price, S. J. (1967). *Canad. J. Chem.* **45**, 575.
39. Thynne, J. C. J. (1961). *Proc. Chem. Soc.* 68.
40. Kerr, J. A. and Trotman-Dickenson, A. F. (1959). *J. Chem. Soc.* 1602.
41. Terry, J. O. and Futrell, J. H. (1967). *Canad. J. Chem.* **45**, 2327.
42. Knox, J. H. (1965). *Combustion and Flame* **9**, 297.
43. Hay, J. M. (1967). *Combustion and Flame* **11**, 83.
44. Russell, G. A. and Bridger, R. F. (1963). *J. Amer. Chem. Soc.* **85**, 3765.
45. Bennett, J. E., Mile, B., Thomas, A. and Ward, B. (1970). *Adv. Phys. Org. Chem.* **8**, 1.
46. Hoare, D. E. and Pearson, G. S. (1964). *Adv. Photochem.* **3**, 121.
47. Hay, J. M. and Hessam, K. (1971). *Combustion and Flame* **16**, 237.
48. Hay, J. M. (1965). *J. Chem. Soc.* 7388.
49. Rust, F. F. and Vaughan, W. F. (1940). *J. Org. Chem.* **5**, 472.
50. Hay, J. M. and Lyon, D. (1970). *Proc. Roy. Soc.* A, **317**, 1.
51. Strausz, O. P. and Gunning, H. E. (1964). *Trans. Faraday Soc.* **60**, 347; Norrish, R. G. W. and Napier, J. M. (1965). *Nature* **208**, 1090.
52. Minkoff, G. J. and Tipper, C. F. H. (1962). *In* "Chemistry of Combustion Reactions", Butterworths, London.
53. Pankratova, V. N., Latyaeva, V. N. and Razuvaev, Zh. (1965). *Obshch. Khim.* **35**, 904.
54. Lewis, B. and von Elbe, G. (1961). "Combustion, Flames and Explosions of Gases", Academic Press, New York; Shtern, V. Ya. (1964). "The Gas Phase Oxidation of Hydrocarbons", Pergamon, Oxford; Tipper, C. F. H. and Minkoff, G. J. (1962). "The Chemistry of Combustion Reactions", Butterworths, London.
55. Semenov, N. N. (1967). *In* "Photochemistry and Reaction Kinetics", Cambridge University Press.
56. Hay, J. M. and Lyon, D. (1970). *Proc. Roy. Soc.* A, **317**, 41.
57. Lewis, B. and von Elbe, G. (1957). "Combustion, Flames and Explosions of Gases". Academic Press, New York.
58. Walsh, A. D. Ninth Symposium (International) on Combustion. (1963). 1064. Academic Press, New York.
59. Freeman, G. R., Danby, C. J. and Hinshelwood, Sir C. N. (1958). *Proc. Roy. Soc.* A, **245**, 28.
60. McMillan, G. R. (1961). *J. Amer. Chem. Soc.* **83**, 3018; Ferguson, J. M. and Philips, L. (1965). *J. Chem. Soc.* 4416; Cox. D. L., Livermore, R. A. and Philips, L. (1966). *J. Chem. Soc. B*, 245.
61. Rice, F. O. and Varnerin, R. E. (1954). *J. Amer. Chem. Soc.* **76**, 324.
62. *See* Gowenlock, B. G. (1965). *Progr. Reaction Kinetics* **3**, 171.
63. Quinn, C. P. personal communication to the author of ref. 62.
64. Norrish, R. G. W. and Pratt, G. L. (1963). *Nature* **197**, 143.
65. Purnell, J. H. and Quinn, C. P. (1967). *In* "Photochemistry and Reaction Kinetics", Cambridge University Press.
66. Peard, M. G., Stubbs, F. J. and Hinshelwood, Sir C. N. (1952). *Proc. Roy. Soc.* A, **214**, 471.
67. Waring, C. E. and Barlow, C. S. (1949). *J. Amer. Chem. Soc.* **71**, 1519.
68. Barnard, J. A. (1957). *Trans. Faraday Soc.* **53**, 1423.
69. Rabinowitch, B. S. and Flowers, M. C. (1964). *Quart Rev.* **18**, 122.

70. Baldwin, R. R., Simmons, R. F. and Walker, R. W. (1966). *Trans. Faraday Soc.* **62**, 2486.

71. Mile, B. (1968). *Angew.* **7**, 507; Bennett, J. E., Mile, B., Thomas, A. and Ward, B. (1970). *Adv. Phys. Org. Chem.* **8**, 1.

72. Bloor, J. E., Brown, A. C. R. and James, D. G. L. (1966). *J. Phys. Chem.* **70**, 2191; *see also* Flannery, J. B. Jr. (1966). *J. Phys. Chem.* **70**, 3707.

73. Noyes, R. M., Applequist, D. E., Benson, S. W., Golden, D. M. and Skell, P. S. (1967). *J. Chem. Phys.* **46**, 1221.

74. Gazith, M. and Szwarc, M. (1957). *J. Amer. Chem. Soc.* **79**, 3339; Feld, M. and Szwarc, M. (1960). *J. Amer. Chem. Soc.* **82**, 3791.

75. Tedder, J. M. and Walton, J. C. (1966). *Trans. Faraday Soc.* **62**, 1859.

76. Heiba, E. A. J. (1966). *J. Org. Chem.* **31**, 776.

77. Shelton, J. R. and Uzelmeier, C. W. (1966). *J. Amer. Chem. Soc.* **88**, 5222.

78. Petersson, G. A. and McLachlan, A. D. (1966). *J. Chem. Phys.* **45**, 628; Atherton, N. N. and Hinchliffe, A. (1967). *Mol. Phys.* **12**, 349; Dixon, W. T. (1965). *Mol. Phys.* **9**, 201.

79. Leone, J. A. and Koski, W. S. (1966). *J. Amer. Chem. Soc.* **88**, 656.

80. Gray, P. and Jones, A. (1965). *Canad. J. Chem.* **43**, 3485.

81. Gray, P. and Jones, A. (1965). *Chem. Comm.* 606.

82. Morris, E. R. and Thynne, J. C. J. (1968). *Trans. Faraday Soc.* **64**, 2124.

83. Gray, P. and Herod, A. A. (1968). *Trans. Faraday Soc.* **64**, 1568.

84. Morris, E. R. and Thynne, J. C. J. (1968). *Trans. Faraday Soc.* **64**, 414.

85. Fessenden, R. W. (1967). *J. Phys. Chem.* **71**, 74.

86. Lin, M. C. and Laidler, K. J. (1968). *Canad. J. Chem.* **46**, 479.

87. Burkley, I. B. and Rebbert, R. E. (1963). *J. Phys. Chem.* **67**, 168; Sanders, W. A. and Rebbert, R. E. (1963). *J. Phys. Chem.* **67**, 170; Wanderlich, F. J. and Rebbert, R. E. (1963). *J. Phys. Chem.* **67**, 1382.

88. *See* Saloman, M. (1964). *Canad. J. Chem.* **42**, 610.

89. Kondrat'ev, V. N. (1965). *Russ. Chem. Rev.* **34**, 893.

90. Fukui, K., Kato, H. and Yonegawa, T. (1961). *Bull. Chem. Soc. Japan* **34**, 1111.

91. Bloor, J. E. and Gartside, S. (1959). *Nature* **184**, 1313.

92. Bennett, J. E., Mile, B., Thomas, A. and Ward, B. (1970). *Adv. Phys. Org. Chem.* **8**, 1; Mile, B. (1968). *Angew.* **7**, 507; *see also* Srinivasan, S. and Carlough, K. H. (1967). *Canad. J. Chem.* **45**, 3209.

93. Fielding, W. and Pritchard, H. O. (1962). *J. Phys. Chem.* **66**, 821; Duncan, F. J. and Trotman-Dickenson, A. F. (1962). *J. Chem. Soc.* 4672.

94. Heiba, E. A. J. and Dessau, R. M. (1967). *J. Amer. Chem. Soc.* **89**, 3772.

95. Walling, C. and Paduva, A. (1961). *J. Amer. Chem. Soc.* **83**, 2207.

96. Garcia Dominguez, J. A. and Trotman-Dickenson, A. F. (1962). *J. Chem. Soc.* 940.

97. Bass, C. D. and Pimentel, G. C. (1961). *J. Amer. Chem. Soc.* **83**, 3754.

98. Kibby, C. L. and Weston, R. E. Jr. (1968). *J. Amer. Chem. Soc.* **90**, 1084.

99. *See* Rebbert, R. E. and Ausloos, P. (1967). *J. Chem. Phys.* **47**, 2849.

100. Doepker, R. C. and Ausloos, P. (1964). *J. Chem. Phys.* **41**, 1865; Rebbert, R. E. and Ausloos, P. (1968). *J. Chem. Phys.* **48**, 306.

101. Doepker, R. C. and Ausloos, P. (1964). *J. Chem. Phys.* **41**, 1865.

102. Rebbert, R. E. and Ausloos, P. (1967). *J. Chem. Phys.* **47**, 2489.

103. Cadman, P. and Simons, J. P. (1966). *Trans. Faraday Soc.* **62**, 631.

104. Hay, J. M. and Norrish, R. G. W. (1965). *Proc. Roy. Soc. A*, **288**, 17.

105. Mitchell, R. C. and Simons, J. P. (1967). *Disc. Faraday Soc.* **44**, 208.

106. Magee, J. L. and Hamill, W. H. (1959). *J. Chem. Phys.* **31**, 1380.
107. Pechukas, P., Light, J. C. and Rankin, C. (1966). *J. Chem. Phys.* **44**, 794; Light, J. C. and Lin, L. (1965). *J. Chem. Phys.* **43**, 3204.
108. Basco, N., James, D. G. L. and James, F. C. (1972). *Inter. J. Chem. Kinetics* **4**, 129.
109. Hase, W. L., Johnson, R. L. and Simons, J. W. (1972). *Inter. J. Chem. Kinetics* **4**, 1.
110. Hesse, C. and Roncin, J. (1970). *Mol. Phys.* **19**, 803.
111. Riley, S. J. and Wilson, K. R. (1972). *Disc. Faraday Soc.* **53**, 132.

CHAPTER 5

THE FORMATION AND REACTIVITY OF RADICALS: POLAR EFFECTS

In the previous chapter, an attempt was made to describe a number of the features of free radical activity using simple concepts and avoiding as far as possible the introduction of polar factors. We shall now consider some of the complications arising from polar effects in free radical reactions, but bearing in mind that such effects can range from a mild perturbation to a completely controlling influence, the treatment is necessarily more superficial than that for non-polar effects; it is also confined to hydrogen transfer reactions where kinetic data are more extensive and reliable. Reactions of radical ions which have been the subject of extensive study[1] will not be considered here.

As we saw in Chapter 3 an empirical relationship can be established measuring the reactivity of an organic free radical by its ability to abstract a hydrogen atom from a series of substrates. These "Polanyi"-type correlations, however, apply only to reactions between a radical and a number of closely analogous substrates, for example the alkanes, and break down when extended to different reactants. Clearly other factors in addition to bond dissociation energy are involved and it is not unreasonable to conclude that at least one of these is the effect of polarity both of radical and of substrate.

It is possible to predict the electronegativities of free radicals on the basis of the group electronegativities given in Table III, Chapter 2. If the hydrogen atom is taken for convenience as the standard, most atoms and groups with configurations unchanged from those in the table will be electrophilic. Alkyl radicals in the tetrahedral configuration fall into this category whilst the planar form with its free electron in the 2p orbital of the trigonal carbon atom is nucleophilic. The nucleophilicity diminishes with increase in the electronegativity of an attached substituent.

Consider the series of plots making up Fig. 1. These give the activation energies of hydrogen atom transfer reactions for a series of atoms and radicals against the differences in the dissociation energies of the bonds being broken and formed. In order to attempt to separate polar effects due to radicals from those attributable to the substrate molecules the full lines are drawn with a slope of -0.5 through alkane substrates, where possible methane, ethane and isobutane (to give the t-butyl radical) these reactions being

135

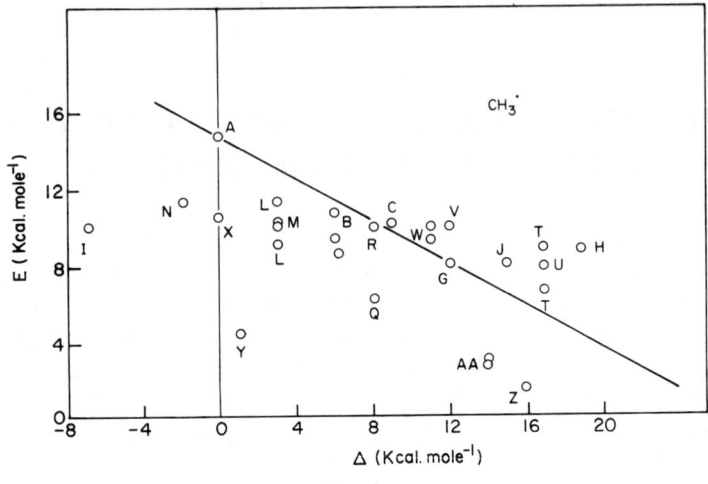

FIG. 1a

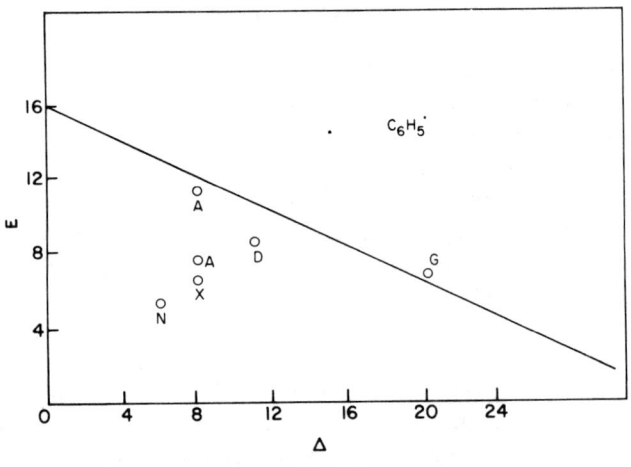

FIG. 1b

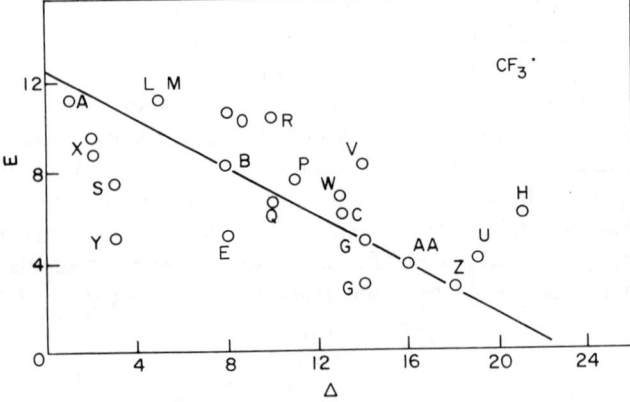

FIG. 1c

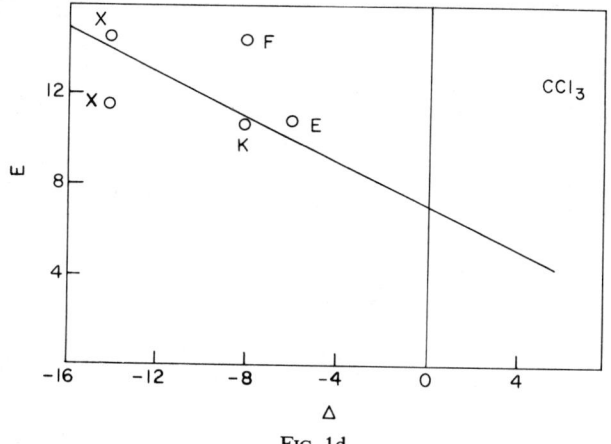

FIG. 1d

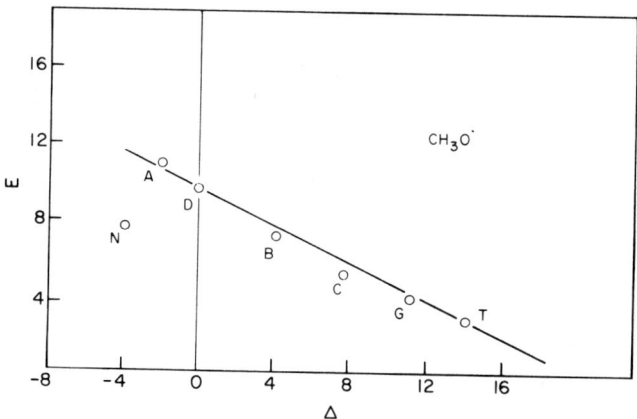

FIG. 1e

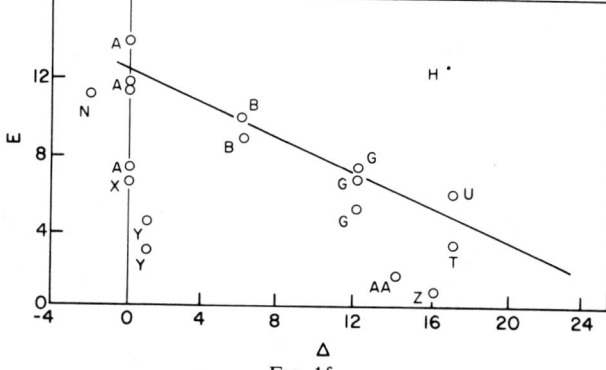

FIG. 1f

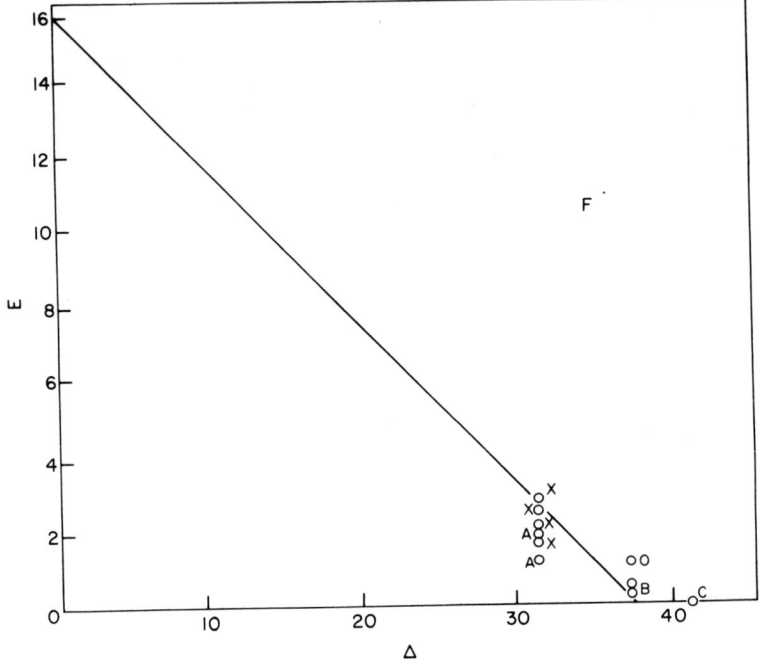

FIG. 1g

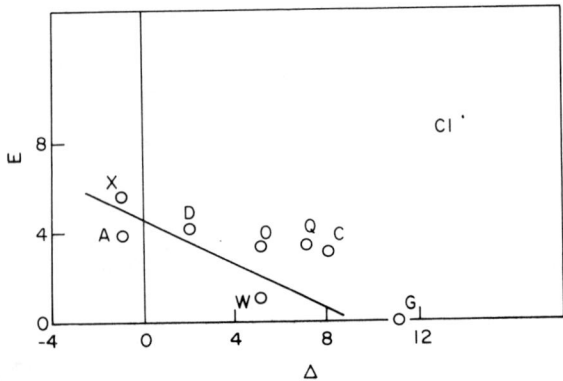

FIG. 1h

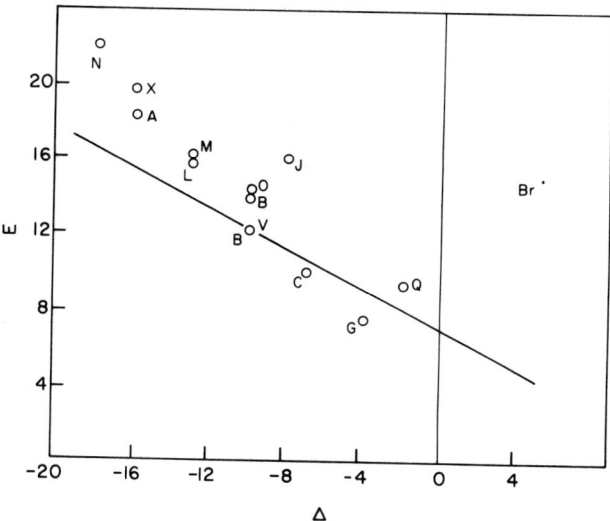

FIG. 1i

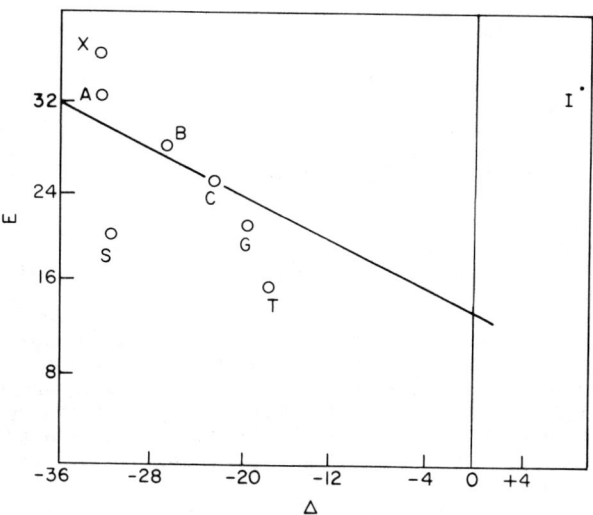

FIG. 1j

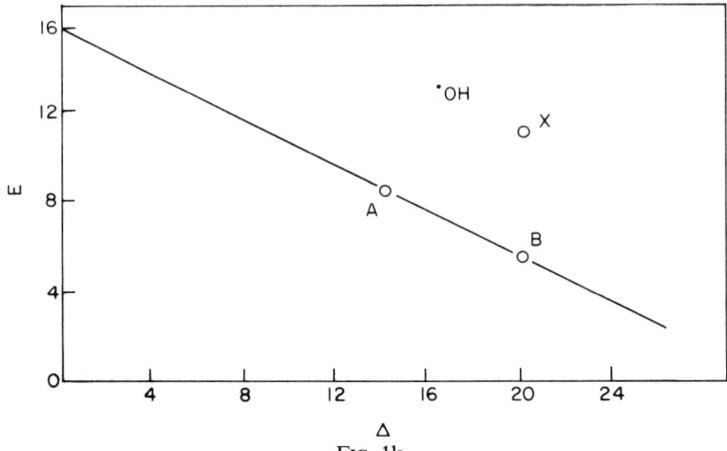

FIG. 1k

A, CH_4; B, C_2H_6; C, C_3H_8; D, cyclo-C_3H_8; E, n-C_4H_{10} (secondary C—H); F, n-C_4H_{10} (primary C—H); G, i-C_4H_{10}; H, $C_6H_5CH_3$; I, C_2H_4; J, CH_3CH=CH_2; K, cyclo-C_6H_{12}; L, CH_3F; M, CH_2F_2; N, CHF_3; O, CH_3Cl; P, CH_2Cl_2; Q, $CHCl_3$; R, CH_3Br; S, CH_3I; T, HCHO; U, CH_3CHO; V, CH_3COCH_3; W, CH_3OCH_3; X, H_2; Y, HCl; Z, HBr; AA, H_2S.

Bond Dissociation Energies from: Benson, S. W. (1965). J. Chem. Ed. 42, 502; Vedenyev, V. I. et al. (1966). "Bond Energies, Ionisation Potentials and Electron Affinities", Arnold, London; Calvert, J. G. and Pitts, J. N. (1966). "Photochemistry", Wiley, New York; Tarr, A. M., Coomber, J. W. and Whittle, E. (1965). Trans. Faraday Soc. 61, 1182; Coomber, J. W. and Whittle, E. (1966). Trans. Faraday Soc. 62, 1553; Walsh, R., Golden, D. M. and Benson, S. W. (1966). J. Amer. Chem. Soc. 88, 650.

(a) Methyl abstraction. A, L, M, N;[a] B, G, H, J, L, Q, U, V, W, X;[b] I, O, R;[c] L, M;[d] Y[e]; AA[f,t]; T; [g]Z;[h] W.[s]

(b) Phenyl abstraction. A, D, G, N, X.[g]

(c) Trifluoromethyl abstraction. A, G, H, O, Q, U, V, X;[i] A, B, L, M, O, P, Q, R;[a] Y, Z;[j] AA;[k] H;[g] W.[u]

(d) Trichloromethyl abstraction. E, F, X;[bb] K;[cc] X.[aa]

(e) Methoxyl abstraction. A, B, C, D, G, T;[q] N.[p]

(f) Hydrogen atom abstraction. A, B, G, X;[g] A, B, G, T, X, Y, Z;[m] N;[j] Y;[n] U;[o] AA.[v]

(g) Fluorine atom abstraction. A, B, O, X;[dd] A, B, X;[l] X.[ee]

(h) Chlorine atom abstraction. A, B, D, G, O, P, Q, X.[l]

(i) Bromine atom abstraction. A, B, L, M, N;[a] C, G, O, Q, R, X;[l] A;[w] B;[x] V.[y]

(j) Iodine atom abstraction. A, B, C, G, U, X.[r]

(k) Hydroxyl abstraction. A, B, X.[z]

[a] Giles, R. D., Quick, L. M. and Whittle, E. (1967). Trans. Faraday Soc. 63, 662; [b] Benson, S. W. (1964). Adv. Photochem. 2, 1; [c] Benson, S. W. (1960). "The Foundations of Chemical Kinetics", McGraw Hill, New York; [d] Pritchard, G. O., Bryant, J. T. and Thommarson, R. L. (1965). J. Phys. Chem. 69, 664; [e] Amphlett, J. C. and Whittle, E. (1966). Trans. Faraday Soc. 62, 1662; [f] Arthur, N. L. and Bell, T. N. (1966). Canad. J. Chem. 44, 1445; [l] Trotman-Dickenson, A. F. (1965). Adv. Free Radical Chemistry 1, 1; [h] Feths, G. C. and Trotman-Dickenson, A. F. (1961). J. Chem. Soc. 3037; [i] Tedder, J. M. and Walton, J. C. (1967). Progr. Reaction Kinetics 4, 37; [j] Amphlett, J. C. and Whittle, E. (1967). Trans. Faraday Soc. 63, 2695; [k] Imai, N. and Toyama, O. (1960). Bull. Chem. Soc. Japan 33, 652; [l] Feths, G. C. and Knox, J. H. (1964). Progr. Reaction Kinetics 2, 1; [m] Thrush, B. A. (1965). Progr. Reaction Kinetics 3, 65; [n] Clyne, M. A. A. and Stedman, D. H. (1966). Trans Faraday Soc. 62, 2164; [o] Frost, W. R. Darwent, B. de B. and Steacie, E. W. R. (1948). J. Chem. Phys. 16, 353; [p] Morris, F. R. and Thynne, J. C. J. (1968). Trans. Faraday Soc. 64, 414; [q] Gray, P., Shaw, R. and Thynne, J. C. J. (1967). Progr. Reaction Kinetics 4, 63; [r] Benson, S. W. and De-

More, W. B. (1965). *Ann. Rev. Phys. Chem.* **16**, 397; [s] Loucks, L. F. (1967). *Canad. J. Chem.* **45**, 2775; [t] Gray, P., Herod, A. A. and Leyshon, I. (1969). *Canad. J. Chem.* **47**, 689; [u] Arthur, N. L., Gray, P. and Herod, A. A. (1969). *Canad. J. Chem.* **47**, 1347; [v] Kurylo, M. J., Peterson, N. C. and Braun, W. (1971). *J. Chem. Phys.* **54**, 943; [w] Gac, N. A., Golden, D. M. and Benson, S. W. (1969). *J. Amer. Chem. Soc.* **91**, 3091; [x] King, K. D., Golden, D. M. and Benson, S. W. (1970). *Trans. Faraday Soc.* **66**, 2794; [y] King, K. D., Golden, D. M. and Benson, S. W. (1970). *J. Amer. Chem. Soc.* **92**, 5541; [z] Avramenko, L. I. and Kolesnikova, R. V. (1964). *Adv. Photochem.* **2**, 25; [aa] Hautecloque, S. (1970). *J. Chim. Phys. Physicochim. Biol.* **67**, 771; [bb] Sidebottom, H. W. Tedder, J. M. and Walton, J. C. (1972). *Inter. J. Chem. Kinetics* **4**, 249; [cc] White, M. L. and Kuntz, R. R. (1971). *Inter. J. Chem. Kinetics* **3**, 127; [dd] Foon, R. and Reid, G. P. (1971). *Trans. Faraday Soc.* **67**, 3513; [ee] Kapralova, G. A., Margolin, A. L. and Chaikin, A. M. (1970). *Kinetics and Catalysis* **11**, 669.

chosen so that the reaction is least perturbed by polar effects. The lines have the equation,

$$E = -0.5\Delta BDE + a$$

where a is the intercept which, since all other factors are essentially being held constant, must be some function of the attacking radical.

A possible function is the electronegativity of the radical and Fig. 2 is a plot of the electronegativities of the relevant radicals against the series of intercepts, a, from Fig. 1. For the majority of the radicals the plot (i) shows a smooth reduction in a with increase in electronegativity of the attacking radical. This trend can readily be explained in terms of the expected transition state for the abstraction of a hydrogen atom from a saturated hydrocarbon,

$$X^{\cdot} + HR \rightleftharpoons X\cdots\underset{\xleftarrow{x}}{H}\cdots R \rightleftharpoons XH + R^{\cdot}$$

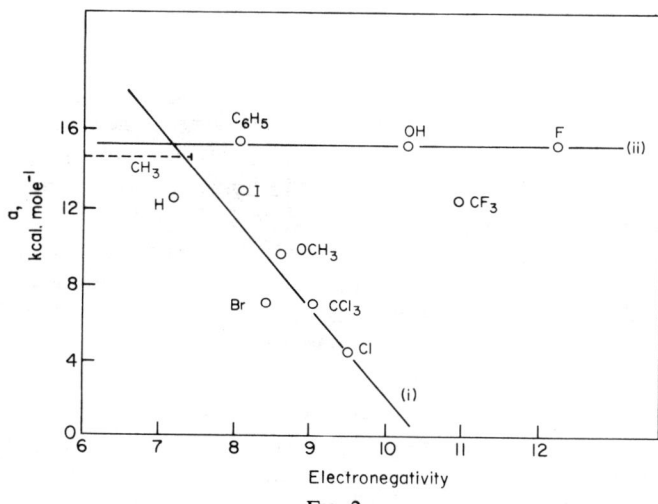

FIG. 2

As we have seen, except under conditions where the resulting radical is in the σ-configuration, the formation of the transition state is accompanied by the flattening of the inchoate radical, the trigonal carbon atom moving towards a planar sp^2 hybridization and the orbital overlapping the hydrogen atom in course of transfer becoming more p in character with an electronegativity less than that of the hydrogen atom. The charge distribution in the transition state is thus towards the hydrogen atom being abstracted (i) and there will therefore be a polar contribution towards the activation energy of abstraction which will therefore decrease with increase in the electronegativity of the attacking radical, as found,

$$\overset{\delta+}{X}\cdots\overset{\delta-}{H}\cdots\overset{|}{\underset{|}{\overset{/}{C}}}\,\delta+ \tag{i}$$

The radicals with the highest electronegativity, however, namely OH˙, CF_3˙ and F˙ do not follow this trend and instead their a values lie along a line, (ii), drawn horizontally from the intercept of the methyl radical where electronegativity effects are least. It would appear that with these highly reactive radicals, electronegativity effects are unimportant and again this is entirely consistent with our picture of radical reactivity if it is the case that, because of the reactivity of these radicals the transition state for the abstraction of a hydrogen atom from a hydrocarbon involves a smaller perturbation of the inchoate alkyl radical or, in other words, the hydrogen atom transfer takes place at greater distances, x, than is the case with the less reactive radicals (ii),

$$X\cdots H\cdots \overset{/}{\underset{\backslash}{C}} \tag{ii}$$

Little or no assistance towards the activation energy is expected from polar effects in such a transition state; indeed, if the alkyl moiety is unperturbed, charge distribution will be away from the hydrogen atom and attack by an electrophilic agent will be discouraged.

Two other points about Fig. 2 seem worthy of mention. First, the electronegativity of the methyl radical cannot be assigned with certainty since during the course of abstracting a hydrogen atom, it is deforming from planar to (ultimately) tetrahedral. Second, the intercept of the phenyl radical appears to place it in the same category as the highly reactive radicals, OH˙, CF_3˙ and F˙ which is reasonable if, due to the highly directional nature of the sp^2 orbital containing the free electron of the phenyl radical, abstraction takes place at greater distances, x, than is the case with the radicals whose intercepts lie on the line (i) in Fig. 2.

Turning now to the substrates, the electronegativity of the moiety attached to the hydrogen atom being abstracted is clearly very important but in view of the scatter of the points in Fig. 1 it is unlikely that we can derive any general relationship. However, one or two observations are worth noting. Decrease in the charge on the hydrogen atom being abstracted, for example by moving from the alkane to the haloalkane series, increases the activation energy for transfer by electrophilic relative to nucleophilic radicals since now the transition state is stabilized in the form,

$$\overset{|\delta-\delta+}{\underset{\diagup\quad\diagdown}{C}}\cdots H\cdots\overset{\diagup}{\underset{|\diagdown}{C}}$$

This effect occurs to such an extent in the abstraction by methyl of a hydrogen atom from acetylene that the activation energy may be less than the difference in the bond dissociation energies.[2] The opposite trend is shown by chlorine atoms in the reaction.[3]

$$Cl^{\cdot} + CHCl_3 \rightarrow HCl + CCl_3^{\cdot}$$

which has a particularly high activation energy. Similar effects are seen in the series of substrates HX where X is H, Cl, Br and SH.

A curious effect is observed in the series CH_4, CH_3F, CH_2F_2 and CHF_3 where the activation energy for abstraction by an electrophilic radical would be expected to rise steadily. But, as the following results show[4] the experimental activation energies for bromine atom attack exhibit a minimum at CH_3F. For comparison the results for methyl abstraction are also given.

RH	$D(R-H)$ kcal mol^{-1}	$Br^{\cdot}(CH_3^{\cdot}) + RH \rightarrow HBr(CH_4) + R^{\cdot}$ E_{Br} kcal mol^{-1}	E_{CH_3} kcal mol^{-1}	% s-character of free electron[5]
CH_4	104	18·3	14·8	0
CH_3F	101	15·8	11·4	$c.$ 0
CH_2F_2	101	16·3	10·2	10
CHF_3	106	22·1	11·4	21

The last column gives the percentage s-character of the resulting radical. We can explain this on the basis of a competition between polar effects and bond dissociation energy differences. On going from CH_4 to CH_3F the activation energy for (electrophilic) bromine atom attack tends to increase due to the decreased charge on the hydrogen atom being abstracted but this is more than offset by the reduction in $D(CH_2F{-}H)$, both effects being caused by the electronegativity of the fluorine atom. As between CH_3F and CH_2F_2 the charge on the hydrogen atom is further reduced but since the resulting radical $CHF_2{}^{\cdot}$ is no longer quite planar it therefore contributes a smaller reorganizational energy to the reaction and the value of the bond dissociation energy is unchanged. The activation energy therefore increases. The last radical, $CF_3{}^{\cdot}$, is pyramidal, the bond dissociation energy is increased and this, together with the further decrease in the charge on the hydrogen atom produces a considerable increase in the activation energy. The activation energy for hydrogen abstraction by the nucleophilic methyl radical shows the expected decrease on going from CH_4 to CH_2F_2 but increases slightly with CHF_3 no doubt due to increase in the bond dissociation energy, $D(CF_3{-}H)$.

Finally we might predict that, should some property of the substrate bring about an increase in the distance, x, in the transition state, the activation energy for hydrogen abstraction will be greater than expected. This appears to be the case with aldehydes where the resulting radicals are σ and substances such as propylene and toluene where the product radicals are strongly stabilized by resonance effects.[6]

In general, therefore, we can say that the activation energy for a radical hydrogen transfer reaction depends upon (a) the difference in dissociation energy between the bond being broken and that being formed, (b) the charge on the hydrogen atom and (c) the structure of the resulting radical.

An interesting sideline of the type of argument set out above is the possibility of deriving information about the structure of radicals. Consider, for example, the ethynyl radical which, as we have seen in Chapter 2 can be written in three ways,

(a)   (b)   (c)

The discussion in Chapter 2, although not conclusive, pointed towards (c) as the most probable structure. As such the ethynyl radical would be expected to be nucleophilic, structures (a) and (b) being electrophilic. The structure (c) is confirmed by evidence from hydrogen abstraction reactions of ethynyl which more readily removes a hydrogen atom from ethylene than from propylene.[7] The hydrogen atoms in the former olefin are expected to carry a lower charge than the methyl hydrogen atoms of propylene, which are assumed to be attacked, because of the greater electronegativity of the sp^2 bonded carbon.

REFERENCES

1. Kaiser, E. T. and Kevan, L. (eds.) (1965). "Radical Ions", Interscience, New York.
2. Drew, C. M. and Gordon, A. S. (1959). *J. Chem. Phys.* **31**, 1417.
3. Feths, G. C. and Knox, J. H. (1964). *Progr. Reaction Kinetics* **2**, 1.
4. Giles, R. D., Quick, L. M. and Whittle, F. (1967). *Trans. Faraday Soc.* **63**, 662.
5. Fessenden, R. W. and Schuler, R. H. (1965). *J. Chem. Phys.* **43**, 2704.
6. Szwarc, M. "The Transition State" (1962). Chem. Soc. Special Publications.
7. Tarr, A. M., Strausz, O. P. and Gunning, H. E. (1966). *Trans. Faraday Soc.* **62**, 1221.

Author Index

147

149

Knudsen, G. A. Jr., 14 (31), *30*
Kolesnikova, R. V., 141
Kominar, R. J., 101 (38), *132*
Konaka, R., 37 (14), *58*
Kondrat'ev, V. N., 122 (89), *133*
Koski, W. S., 118 (79), *133*
Kozyrev, B. M., 32 (4g), *57*
Krech, M., 94 (19), *131*
Kreilick, R. W., 40 (25), 41, *58*
Krusic, P. J., 39
Kubin, R. F., 92 (15), *131*
Kuitu, L., 81 (34), *85*
Kuntz, R. R., 141
Kurylo, M. J., 141
Kutschke, K. O., 119
Kyle, L. M., 48 (42), *58*

L

Laidler, K. J., 5, (3), *30*, 90, 91 (11), (12), 94 (19), 102, 121 (86), *131, 133*
Lankamp, H., 18 (42), *31*, 44 (30), *58*
Latta, B. M., 27 (60), *31*
Latyaeva, V. N., 105 (53), *132*
Lau, P. W., 38 (17), 39, *58*
Lazdius, D., 36 (9), *57*
Legett, C., 91 (8), *131*
Leone, J. A., 118 (79), *133*
LeRoy, D. J., 13 (25), *30*
Levy, D. H., 36 (8), *57*
Lewis, B., 106 (54), 108 (57), *132*
Leyshon, I., 141
Light, J. C., 130 (107), *134*
Lin, L., 130 (107), *134*
Lin, M. C., 91 (11), (12), (13), 121 (86), *131, 133*
Lin, W. C., 38 (17), 39, *58*
Lin, W. E., 49 (43), *58*
Linnett, J. W., 8 (11), 26 (55), (56), *30, 31*, 100 (36), *132*
Lissi, E. A., 93 (16), 94 (17), 102, *131*
Lifshitz, C., 81, 82 (36), *86*
Liu, M. T. H., 94 (19), *131*
Livermore, R. A., 91 (6), 109 (60), *131, 132*
Lossing, F. P., 12 (22), *30*, 96
Loucks, L. F., 94 (19), *131*, 141
Lown, I. W., 21 (48), *31*
Lyon, D., 53 (47), *58*, 104 (50), 107 (56), 123 (50), 127 (50), *132*

M

McConnell, H. M., 36 (6), 40 (24), 54 (6), 57, *58*
McDowell, C. A., 64 (6), *85*
McKellar, J. F., 100 (35), *131*
McLachlan, A. D., 32 (4l), *57*, 117 (78), *133*
MacLean, C., 18 (42), *31*, 40 (23), 44 (30), *58*
McMillan, G. R., 102, 109 (60), *132*
MacMillan, J., 9 (15), *30*
McNesby, J. R., 16 (37), *30*, 98 (31), 123, *131*
Magee, J. L., 100, 130 (106), *134*
Maki, A. H., 16 (35), *30*, 37 (14), *57*
Marcus, 93
Margolin, A. L., 141
Marta, F., 109
Martens, G., 97 (29), *131*
Marvel, C. S., 17 (40), *30, 31*
Mayo, F. R., 20 (43), *31*
Melchior, M. T., 37 (14), *57*
Melville, H. W., 97 (28), *131*
Menczel, J. H., 94 (19), *131*
Metcalfe, E. L., 89
Meyers, L., 97 (29), *131*
Mile, B., 39, 40 (28), *58*, 103 (45), 112 (71), 123 (92), *132, 133*
Minkoff, G. J., 105 (52), 106 (54), *132*
Mitchell, R. C., 55, 130 (105), *133*
Miyoshi, M., 119
Molyneux, P., 88 (3), *131*
Morris, E. R., 121 (82), (84), *133*, 140
Mortimer, C. T., 74 (24), 83, *85*
Moss, S. J., 27 (59), *31*
Mueller, M. B., 17 (40), *30, 31*
Mulcahy, M. F. R., 94 (19), *131*
Mutter, W. E., 109

N

Nakazaki, M., 37 (11), *57*
Napier, J. M., 104 (51), *132*
Nauta, W. Th., 18 (42), *31*, 44 (30), *58*
Neumann, M. G., 93 (16), 102, *131*
Ni Yao Tang, 75 (26), *85*
Norman, R. O. C., 34, 38 (19), (21), 39, 47 (39), *58*

Subject Index

A

a (hyperfine coupling constant), 34–38, 47

Abstraction reactions, 18, 67, 69–71, 119–128, 130, 135–144

Acetyl radical,
  decomposition, 92
  i.r. spectrum of, 40
  structure, 55
  +NO, 109
  +O_2, 103

Acetylene,
  combustion of, 104, 123, 127, 130
  dissociation, 72

Activation energy,
  of abstraction, 122–123
  of addition, 114–115
  of decomposition, 111
  of dissociation, 43–44, 87

Acyl radicals,
  e.s.r. of, 39
  structure of, 55
  +NO, 109

Addition reactions, 18–21, 112–118

Alkoxy radicals,
  combination of, 100
  decomposition of, 7, 90–91
  disproportionation of, 100
  methyl shift in, 14
  +NO, 109

Alkyl radicals,
  combination of, 16–18
  decomposition of, 15, 91–96
  disproportionation of, 16–17
  isomerization of, 12–14
  structure of, 68
  +NO, 109–111
  +O_2, 102–108, 128–130

Allyl radicals,
  resonance in, 47, 81

Aromatic substitution, 21–23, 116–118
  aryl radicals by, 21–23
  inductive effects in, 22
  mechanism of, 22, 118
  mesomeric effects in, 117
  methyl radical by, 22–23
  steric effects in, 21, 117–118

Arrhenius factor (A),
  for abstraction, 119–122
  for addition, 112–115
  for decomposition, 42–44, 87–96, 111
  for disproportionation, 100–101

Aryloxy radicals,
  disproportionation of, 17
  hydrogen atom transfer in, 14

B

b.d.e., 42, 59–84

Benzoyl,
  e.s.r. spectrum of, 39
  structure, 55

Benzyl,
  resonance in, 24, 46–47, 101

Bimolecular reactions, 7–8, 11, 16–24
  pressure effect in, 8

Bond energies, 59–84
  effect on radical stability, 49–50, 52–56
  experimental methods to obtain, 61–71

Bromine atom,
  abstraction reactions, 139, 141, 143
  addition reactions, 114–116

t-butyl,
  σ-radical, 43, 95

154

155

formation, 2–4
reactions, 11–24
reactivity, 24–29
  resonance on, 24–29
  steric effects, 28–29
resonance (calculated), 80–82, 84
solvent interactions, 10–11
structure, 44–57
+NO, 108–111
+O_2, 102–108

G

g-factor, 32

H

Heats of atomisation, 77–80, 81
Hexa-aryl ethanes, 27, 29
cyclo-Hexadienyl radical,
  resonance energy, 81
  ring strain in, 117–118
Hexene,
  as inhibitor, 109
Hot radicals, 8, 95, 128–130
  quenched by inert gas, 130
Hot recoil tritium atoms,
  in dissociation energy measurement,
    67–68
Hydrazine decomposition, 89
Hydrazyl radicals,
  resonance, 28
Hydrocarbon combustion, 106–108
  cool-flames in, 106, 108
  degenerate branching in, 106–107
  mechanism, 107
  negative temperature coefficient in,
    106, 107–108
Hydrocarbon pyrolysis, 109–111
  inhibition of, 109–111, 112
  mechanism, 111
Hydrogen atoms,
  abstraction by, 135–144
  addition of, 4, 27
    polar effects in, 27
Hydroxyl radicals,
  abstraction by, 140, 141, 142

Hydroxyalkyl radicals,
  e.s.r., 38, 121
  from abstraction reactions, 121
  structure, 55, 121
Hyperconjugation, 47, 84
  by e.s.r., 36

I

i.b.d.e., 42, 73–84
  calculation, 73–77, 81–82
  definition, 42, 73
  tables, 76, 78, 82
Imino radicals,
  resonance, 28
Inhibition, 108–111
1-4 Interaction parameters, 79, 84
Iodine atom,
  abstraction by, 139, 141
  addition of, 113
i.r. spectra of radicals, 40, 44, 54, 55
Isomerization reactions, 11–15, 124–128

L

Liquid phase, 9–11
  diffusion in, 10
  wall effects in, 9

M

McConnell equation, 36–37, 54
Methanol,
  b.d.e., 72
  photochemical decomposition, 63–64
Methoxy radical,
  abstraction by, 137, 141
  structure of, 89
Methyl radical,
  abstraction by, 119–120, 136, 141, 143
    at low temperatures, 123
  addition of, 27, 113, 114–115
  bond energies in, 50
  e.s.r., 44
  inversion energy of, 44–46
  i.r. ot, 44
  structure of, 44–46, 128